Long-Term Performance Prediction for PE Pipes

Awwa Research Foundation

About the Awwa Research Foundation

The Awwa Research Foundation (AwwaRF) is a member-supported, international, nonprofit organization that sponsors research to enable water utilities, public health agencies, and other professionals to provide safe and affordable drinking water to consumers.

The Foundation's mission is to advance the science of water to improve the quality of life. To achieve this mission, the Foundation sponsors studies on all aspects of drinking water, including supply and resources, treatment, monitoring and analysis, distribution, management, and health effects. Funding for research is provided primarily by subscription payments from approximately 1,000 utilities, consulting firms, and manufacturers in North America and abroad. Additional funding comes from collaborative partnerships with other national and international organizations, allowing for resources to be leveraged, expertise to be shared, and broad-based knowledge to be developed and disseminated. Government funding serves as a third source of research dollars.

From its headquarters in Denver, Colorado, the Foundation's staff directs and supports the efforts of more than 800 volunteers who serve on the board of trustees and various committees. These volunteers represent many facets of the water industry, and contribute their expertise to select and monitor research studies that benefit the entire drinking water community.

The results of research are disseminated through a number of channels, including reports, the Web site, conferences, and periodicals.

For subscribers, the Foundation serves as a cooperative program in which water suppliers unite to pool their resources. By applying Foundation research findings, these water suppliers can save substantial costs and stay on the leading edge of drinking water science and technology. Since its inception, AwwaRF has supplied the water community with more than $300 million in applied research.

More information about the Foundation and how to become a subscriber is available on the Web at **www.awwarf.org**.

Long-Term Performance Prediction for PE Pipes

Prepared by:
Paul Davis, **Stewart Burn**, **Scott Gould**, **Mark Cardy**, **Grace Tjandraatmadja**, and **Paul Sadler**
Commonwealth Scientific & Industrial Research Organisation (CSIRO)
Australia

Jointly sponsored by:
Awwa Research Foundation
6666 West Quincy Avenue, Denver, CO 80235-3098

and

Commonwealth Scientific & Industrial Research Organisation (CSIRO)
Limestone Avenue, Canberra, Australia

Published by:

Awwa Research Foundation

American Water Works Association

IWA Publishing

DISCLAIMER

This study was jointly funded by the Awwa Research Foundation (AwwaRF) and the Commonwealth Scientific & Industrial Research Organisation (CSIRO). AwwaRF and CSIRO assume no responsibility for the content of the research study reported in this publication or for the opinions or statements of fact expressed in the report. The mention of trade names for commercial products does not represent or imply the approval or endorsement of AwwaRF and CSIRO. This report is presented solely for informational purposes.

CONTENTS

TABLES

FIGURES

FOREWORD

The Awwa Research Foundation (AwwaRF) is a nonprofit corporation that is dedicated to the implementation of a research effort to help utilities respond to regulatory requirements and traditional high-priority concerns of the industry. The research agenda is developed through a process of consultation with subscribers and drinking water professionals. Under the umbrella of a Strategic Research Plan, the Research Advisory Council prioritizes the suggested projects based upon current and future needs, applicability, and past work; the recommendations are forwarded to the Board of Trustees for final selection. The foundation also sponsors research projects through an unsolicited proposal process; the Collaborative Research, Research Applications, and Tailored Collaboration programs; and various joint research efforts with organizations such as the U.S. Environmental Protection Agency, the U.S. Bureau of Reclamation, and the Association of California Water Agencies.

This publication is a result of one of these sponsored studies, and it is hoped that its findings will be applied in communities throughout the world. The following report serves not only as a means of communicating the results of the water industry's centralized research program but also as a tool to enlist the further support of the nonmember utilities and individuals.

Projects are managed closely from their inception to the final report by the foundation's staff and large cadre of volunteers who willingly contribute their time and expertise. The foundation serves a planning and management function and awards contracts to other institutions such as water utilities, universities, and engineering firms. The funding for this research effort comes primarily from the Subscription Program, through which water utilities subscribe to the research program and make an annual payment proportionate to the volume of water they deliver and consultants and manufacturers subscribe based on their annual billings. The program offers a cost-effective and fair method for funding research in the public interest.

A broad spectrum of water supply issues is addressed by the foundation's research agenda: resources, treatment and operations, distribution and storage, water quality and analysis, toxicology, economics, and management. The ultimate purpose of the coordinated effort is to assist water suppliers to provide the highest possible quality of water economically and reliably. The true benefits are realized when the results are implemented at the utility level. The foundation's trustees are pleased to offer this publication as a contribution toward that end.

David E. Rager
Chair, Board of Trustees
Awwa Research Foundation

Robert C. Renner, P.E.
Executive Director
Awwa Research Foundation

ACKNOWLEDGMENTS

The authors of this report are indebted to the following water utilities and companies for their cooperation and participation in this project:

Stephen Boros, Plastics Pipe Institute, Washington DC, USA
Camille Rubeiz, Plastics Pipe Institute, Washington DC, USA
Predrag Micic, Qenos, Victoria, Australia
Mike Stahmer, Vinidex Pty Ltd, Victoria, Australia
Lee Mizell, Performance Pipe, Texas, USA
Jan Dwiggins, Lamson and Sessions, Missouri, USA
Stanley Mruk, Consultant, USA

Bret Russell, Louisville Water Company, Kentucky, USA
George Flegal, United Water Toms River, New Jersey, USA
Keith Browning, Orlando Utility Commission, Florida, USA
Steve Porter, Greenville Utilities Commission, North Carolina, USA
Peter Fortin, City of Norfolk, Norfolk, Virginia, USA
Dan Schaefer, City of Cincinnati Water Works, Ohio, USA
George De Bono, City of Daly City, California, USA
Jim McNerney, Seattle Water, Washington, USA
Dan Smith, Water One, Kansas, USA
Russell Holmes, City of Ottawa, Ontario, Canada
Doug Seargeant, Epcor Water Services, Edmonton, Alberta, Canada

Greg Moore, SA Water, South Australia, Australia
Alan Watts, South East Water, Victoria, Australia
Farshad Ibrahimi, City West Water, Victoria, Australia
Bernard Mitchell, Sydney Water, New South Wales, Australia
Chris Wight, Yarra Valley Water, Victoria, Australia
Jayam Tennakoon, Ipswich Water, Queensland, Australia

The authors would like to thank the following AwwaRF Staff and Project PAC members for their support and constructive contribution to the project:

Jian Zhang, AwwaRF
Ata Ciechanowski, NSF International,
Sean Morefield, US Army Engineer Research and Dev.
Kanwal Oberoi, Charleston Comm. of Public Works

In addition, the authors would like to thank United Kingdom Water Industry Research (UKWIR) for kindly providing a national database of pipe performance characteristics for the United Kingdom Industry.

EXECUTIVE SUMMARY

In Europe, Polyethylene (PE) pipes have been used for water transportation since the 1950s and for gas transportation since the 1960s. In Australia, PE pipes have been used since the late 1970s. The experience in Europe and the UK is that PE exhibits the lowest failure rates of those pipe materials in use. The majority of reported failures are related to third party damage or poor preparation and contamination of joints and do not correspond to failure of PE pipes under normal service loading conditions. In North America, PE has been used for water supply since the late 1980s and for gas transportation since 1965. While PE is currently used on a relatively small scale by the US water industry, it is used almost exclusively by the US gas industry and more than 90% of all new gas services and small diameter mains installed in the US today are High Density Polyethylene pipe.

CONCLUSIONS AND RECOMMENDATIONS

Current Field Performance of Australian, UK, and US PE Water Pipes

In an attempt to assess the current field performance of PE water pipes, historical failure data was requested from a total of 87 water utilities from the Australia, the UK and the US. While actual failure data was supplied in some cases, the majority of information supplied was anecdotal. A further limitation in interpreting the data was the accuracy of failure descriptions currently used by water utilities. For example, although some water utilities reported that circumferential fractures occurred in PE pipes, their high flexibility should allow for large lateral displacements to be accommodated without this type of fracture occurring. Industry expert opinion in the UK suggests that pipe failures that are described as "fracture" do not actually occur under normal service conditions and are in fact attributed to inadequate installation practice. Poor pipe bedding can result in point loading at the pipe outer surface, which in turn creates localized bending stresses in the pipe wall. For older PE materials with relatively low slow crack growth resistance, this may accelerate crack growth from defects in the pipe wall and result in fracture failure. However, these failures reflect poor pipe installation practice rather than the performance of the PE pipe material under normal service loads. Similarly, failures reported to have occurred in joints and fittings are thought to be the result of poor preparation and jointing practice rather than failure of the joint in normal service. Industry expert opinion suggests that these failures can be avoided by following recommended industry guidelines for jointing and as such, give a false representation of failure rates in PE pipes under normal operating conditions.

However, regardless of data quality issues, recorded failure rate data from the UK indicates an average failure rate of 3.2 per 100km/per year for PE pipes, which is still significantly lower than other pipe materials in use. In Australia, slightly higher average failure rates between 5.9 and 21.2 per 100km/per year were reported, but it should be noted that the length of PE pipe in use (and hence the accuracy of recorded failure rates) will be lower than that used in the UK. While no quantitative failure data was supplied by US water utilities, anecdotal evidence describing fracture failures in small diameter PE service connections was given. However, it should be noted that these older PE materials exhibited significantly lower resistance to slow crack growth in comparison to the current generation of PE materials in use. US industry expert opinion indicates that failures in older PE pipes were also influenced by:

- Point load impingement in "stony" soils
- Excessive bending from poor installation and differential soil settlement
- Sharp changes in geometry at improperly dimensioned insert fittings
- One particular batch of PE pipe produced from a single supplier in 1974 which was gradually embrittled under UV due to inadequate stabilizer levels.

Therefore, (as with the Australian and UK performance data), these failures in early PE grades do not represent the current situation of PE materials with increased slow crack growth resistance. Those US water utilities using PE for larger diameter water supply pipe reported zero failures to date. While this may reflect the performance of newer PE materials with increased slow crack growth resistance, it should also be noted that these pipes are relatively young (less than 10 years old).

Failure Mechanisms in PE Pipe Systems Observed in Service

If surface preparation and installation are poor, electrofusion fittings can fail in PE pipe networks. Photographs from field failures indicate that a common failure mode is the expansion of moisture between the coupling and pipe outer surface. Alternatively, failure to properly clean and scrape the pipe surface to be joined can result in contamination of the joint and early failure. It should be noted however, that these failure modes reflect the quality of jointing practices rather than performance of the PE pipe system under normal operating conditions. These kind of fitting failures are avoidable and with proper preparation and installation procedures, reliable heat fusion joints are obtained.

Depending on the PE material, slow crack growth (SCG) failures can occur in some older PE pipes in service. The SCG mechanism occurs by the stepwise growth and failure of a fibrillated zone (or "craze") which forms at the tip of a defect in the pipe wall. Such defects can be inherent (from the pipe manufacturing process) or incurred by damage during pipe transport and installation. The SCG resistance of a particular material is governed by the resistance to extension and separation of the crack tip craze under static tensile stress. In contrast to older materials, new generation PE materials form large crack tip crazes which resist separation and exhibit significantly increased resistance to SCG failure. Consequently, occurrences of SCG failure in service for current PE materials are extremely rare.

Performance Standards for PE Pipes

In the US, performance standards PE specify material property requirements such as density, melt index, flexural modulus, tensile yield strength and slow crack growth resistance. In contrast, international and Australian standards specify requirements for weathering resistance and slow crack growth resistance for a material when manufactured in pipe form. Material properties such as density, yield strength and flexural modulus are not specified. It is proposed that certain important material criteria are not currently included in the US standards and should be addressed. For example, carbon black sizing and weathering resistance are issues that may influence pipe performance.

A comparison of US and international standards indicate that methods to assess long-term hydrostatic strength of PE pipes are similar. Both subject a series of pipe samples to different hydrostatic pressures and record the time to failure. A long term hydrostatic strength is assigned based on the extrapolation of this data to 50 years. The main difference between methods is in how the transition from ductile rupture failure to slow crack growth failure is handled for this extrapolation. This transition in failure mode produces a down-turn in the gradient of best fit lines to hydrostatic stress vs. time data. The ISO method accounts for data in both ductile and slow crack growth failure regimes and requires a relatively large number of data points. In contrast, the ASTM method excludes data in the slow crack growth regime and a separate procedure is used to check for slow crack growth failures in long term data. However, for modern PE materials with increased resistance to SCG, ambient temperature stress-rupture behavior follows a straight line demonstrating ductile rupture through at least the 50-year intercept. For these materials, both ASTM and ISO methods forecast essentially the same long-term strength.

In designing for pressure transients (or surge) events in PE pipes, US standards incorporate a surge allowance into the pressure rating of the pipe. This factor of 0.5 effectively reduces the pressure class by 50% with the corresponding increases in wall thickness and costs incurred by the water utilities. It is proposed that the practice of surge allowance be examined in more detail since other researchers have reported that polymer pipes can become stiffer under higher rate pressurization and may actually resist higher pressures. It is proposed that pipe re-rating to design against fatigue failure from recurring surge pressures should be addressed in US standards. Although they are not formally included in performance standards, UK and Australian industry guidelines for fatigue de-rating are available that account for surge amplitude, frequency and the required pipe lifetime.

ASTM standards use a notched coupon specimen that is subjected to constant load at elevated temperature to assess slow crack growth resistance of PE pipe systems. In comparison, international and Australian standards require slow crack growth resistance to be demonstrated on notched pipe specimens. Evidence from other researchers indicates that notched tensile coupon specimens provide a more severe test than the likely service conditions for a pipe and restrict the crack tip blunting mechanisms which are likely to occur in circumstances of low notch tip constraint (i.e., surface notches in pipes). Consequently, while notched tensile specimens define the inherent SCG resistance of a particular material in a shorter test time, they should not be used to predict actual pipe performance.

The resistance of PE resins to thermal and UV exposure depends on the stabilization system used, and the level of stabilizers incorporated. International and Australian standards specify a minimum Oxygen Induction Time for both material and manufactured pipes to ensure that adequate heat stabilizers are incorporated into the resin and that these stabilizers have not been fully consumed during pipe extrusion. However, US standards only place a heat stabilizer performance requirement on the base material and have no requirement for manufactured pipe specimens. It is proposed that the different test methods adopted by USA and ISO standards should be compared to ensure that the performance requirements are aligned with each other.

Installation and jointing practices influence the service life of PE pipe. Standard practices and guidelines for fusion jointing are available in the US and it is recommended that water utilities ensure that PE piping is installed in accordance with these guidelines. In other countries such as Australia, national training programs and short-term technical courses have been established for PE installations. In some cases, water utilities require installation contractors to have undergone training before they will be awarded a contract. It is recommended that this practice be adopted in the US and water utilities use contractors with prescribed pipe installation and pipe jointing qualifications.

Accelerated Test Methods for Slow Crack Growth Resistance of PE Water Pipes

Hydrostatic pressure testing of PE materials usually produces a ductile failure mode at ambient temperature. However, at elevated temperatures, a transition to a brittle-like slow crack growth (SCG) failure mode can be observed (depending on the material). The early onset of this transition indicates a material with relatively low SCG resistance. To assess the resistance of a particular material to SCG, elevated temperature testing of notched specimens is widely used in US and International performance standards. In general, each method specifies an applied load and SCG resistance is quantified based on the recorded failure time. In general, the performance of older PE materials can be assessed within practical time frames and ranked consistently between the different test methods available. However, some evidence suggests that material microstructure and crack growth mechanisms at elevated temperatures may depart from those experienced by a pipe in service at ambient temperature. Therefore, while elevated temperature SCG tests provide an effective material ranking tool, results should not be used to predict service lifetime without careful consideration of this departure in crack growth mechanism.

While the SCG ranking of older PE materials can be readily assessed using notched specimens at elevated temperatures, the significantly improved SCG resistance of new PE materials means that test times are impractically long. Recent advances use cyclic loading to accelerate the formation and failure of the crack tip craze zone. While evidence again suggests that the mechanism of crack growth differs from that observed for pipes in service, qualitative ranking of new materials with high SCG resistance is now possible in practical time frames. A new test method has also recently been developed which isolates the craze failure process in PE materials. Initial tests indicate that it can be used for material ranking and, with further development, service life prediction.

Long-Term Performance Prediction Methods for PE Water Pipes

In general, approaches to failure prediction in PE pipes can be divided into empirical and theoretical methods. Empirical methods rely on experimental data from fracture property assessments or hydraulic pressure testing at different applied stresses and temperatures. Based on this data, failure times from elevated temperature testing (which can be obtained in practical time frames) can be transformed to give corresponding failure times at ambient temperature. While these techniques are useful, evidence in the literature suggests that the crack growth mechanisms observed at ambient and elevated temperatures differ. However, sufficient correlation can exist to provide qualitative correlations with observed field performance. One issue with empirical approaches is the increased testing time that is required for new PE materials with improved slow crack growth resistance. Even at elevated temperatures, the performance of these materials is such that test times are impractical.

An alternative to empirical approaches is to develop theoretical lifetime prediction models. For example, fracture mechanics theory uses the concept of a single parameter (the Stress Intensity Factor for example) that characterizes the stress field at the tip of a crack in a PE pipe. In Linear Elastic Fracture Mechanics (LEFM) theory, the dependence of crack growth rate on this parameter can be determined from laboratory tests on small coupon samples and then transferred through a geometrical correction factor to the actual geometry of a flawed pipe in service. However, while this approach may apply to older PE materials, new PE materials can form large crack tip craze zones and the associated plastic deformation within this zone significantly alters the local stress state at the crack tip. Since the extent of this plastic deformation depends on the size and geometry of the structure under analysis, single parameter fracture mechanics predictions for new PE materials may not be transferable from small test geometries to larger pipes in service.

However, previous research has demonstrated that the time dependence of craze failure stress in PE materials can be measured experimentally using circumferentially deep notched tensile (CDNT) specimens. Using the theoretical concept of a reference stress in visco-elastic materials, an empirical relationship between applied craze stress and failure time can then be used to predict the time to crack initiation in more complex test geometries.

Fracture Properties of Australian and US PE Pipes

In order to develop a failure model for PE pipes in service, the applicability of single parameter fracture mechanics must be assessed for the current generation of PE materials in use. Based on an experimental comparison of ductile rupture and measured craze failure stress, it was concluded that single parameter fracture mechanics was *invalid* for PE materials in the US and Australia, and could not be used as the basis for service life modelling.

As an alternative, the Circumferentially Deep Notched Tensile (CDNT) test was used to isolate the craze separation process and directly measure the relationship between applied craze stress and failure time for different US and Australian PE materials. For comparison, tests were also conducted on a sample extracted from a small diameter PE service connection that had failed in the US by point loading and fracture. Although the precise age of this sample was unknown, it is thought to be more than 10 years old and manufactured from a previous generation PE material with relatively low resistance to slow crack growth. A description of each sample tested in shown in the following table.

Sample #	Origin
PE100 (AUS)	Australian, as produced
PE80B (AUS)	Australian, as produced
PE80C (AUS)	Australian, as produced
PE3408 (US)	US, as produced
PE FIELD (US)	US, field failure > 10 years old

Under a static applied tensile load, PE100 (AUS), PE80B (AUS) and PE3408 (US) all exhibited relatively high craze strengths over time, when compared to PE80C (AUS). This indicates that the crack tip craze zones which can form in these materials are more resistant to separation, and indicate a superior SCG resistance. As expected, the lowest performance was observed in PE FIELD (US), with an initially low craze strength, which decreased much faster than the other samples tested. This reflects the poor SCG resistance of the older PE material in comparison to currently available material grades.

Developing a Failure Model for Australian and US PE Pipes

Having determined craze strength vs. failure time characteristics for PE materials, simplified models were developed to predict failure times for PE pipes under combined internal pressure and deflection loads. It was shown that models based on Fracture Mechanics theory grossly over-predicted the time to failure in all tests, reinforcing the conclusion that single parameter fracture mechanics cannot be applied to these materials.

Model predictions based on simple net-section collapse of the remaining un-cracked ligament agree well with short-term tests, and an examination of failed samples indicated that failure occurred by yielding, drawing and rupture in practice. However, for longer term tests, the best agreement was obtained from a model that uses the experimentally determined empirical relationship between applied craze stress and failure time. Examination of failed samples indicated that, although failure ultimately occurred by rupture of drawn material ahead of the crack tip, some evidence of craze formation and separation was also visible. This suggests that the craze strength-based model is the best approach to adopt for these materials

Estimating Failure Rates in PE Water Pipe Networks

For asset management purposes, water utilities require the *probability* of pipe failure to be quantified. When combined with cost consequences of failure, this probability allows the level of risk associated with an asset to be defined. In turn, this can be used to schedule future replacements at the most cost-effective time. Alternatively, if a set of assets in a network can be allowed to fail without incurring too severe consequences (i.e., non-critical reticulation mains fro example), then a water utility requires an estimate of the average failure rate in these assets (i.e., number of failures per 100km/per year) that is expected in the future. When combined with pipe replacement costs, this knowledge allows replacement budgets to be set over the required planning period. However, current long-term failure prediction models for PE pipes assume well-defined and quantifiable loading conditions. As such, they do not allow extrapolation to the network-wide scale through the incorporation of uncertainty and probabilistic modeling.

In order to extend models to estimate failure probability, the uncertainty associated with failure mechanisms must be accounted for. To account for uncertainty in the pipe fracture process, a Monte Carlo simulation model was proposed for PE pipelines. Based on the craze separation and net section collapse models discussed above, Monte Carlo simulation was used to generate a hypothetical pipeline, comprised of a series of independent pipe segments. The maximum inherent defect size in each pipe segment is treated as a stochastic variable, which is represented by the Weibull probability distribution function. A maximum inherent defect size is randomly assigned to each segment and service lifetimes are simulated until the criterion for craze separation and/or net section collapse is satisfied in each pipe segment. The end result of the Monte Carlo simulation is a set of predicted lifetimes which can be subsequently used to estimate failure probability or network wide average failure rates.

To benchmark the model, actual historical failure rates in PE pipes from a set of 17 UK water utilities were compared to predicted failure rates for 110 SDR 11 PE pipes manufactured from Australian PE materials. Simulated curves showed reasonable agreement within the 95% confidence limits for actual failure rates. In comparison to older PE materials, model predictions for new Australian and US PE materials exhibited significantly lower simulated average failure rates. This superior performance is attributed to the increased craze separation resistance and ductile fracture resistance exhibited by these PE materials. These extremely low forecasted average failure rates confirm expectations for currently available PE materials with improved slow crack growth resistance. They are also in good qualitative agreement with the extended predicted service lifetimes based on more empirical approaches such as elevated temperature testing. However, while the failure prediction model is a potentially useful asset management tool for water utilities, simulated failure rates cannot be validated until sufficient historical failure data is available for these new PE materials.

FUTURE RESEARCH

Several areas are identified for future development of the Monte Carlo simulation model.

Inherent Defect Size

It is essential that pipe samples from USA/Canadian field failures are exhumed and their fracture surfaces examined. While an initial examination under optical microscope will reveal the origin of crack initiation, SEM should be used to quantify the inherent defect size. It is also recommended that defect size databases are developed for different pipe diameters. This will allow separate defect probability distributions to be developed for different pipe sizes.

Loading Conditions

The failure model requires the applied circumferential stresses due to internal pressure, external loading, and residual stress as input. While internal pressure and residual stress may be well defined, external loading is idealized in the model as a soil weight and a surface load acting above a buried pipe. A more realistic model should account for external loads generated from differential soil settlement, superimposed loads, and thermal expansion effects. In general, stresses in the pipe wall will also relax over time under a constant load and this should be incorporated into the model. Perhaps more importantly, the response of a buried PE pipe to magnified localized stresses such as those arising from point loading from poor installation should be addressed. Since the Monte Carlo simulation method allows for a great deal of complexity in the failure models, more accurate descriptions of the failure process can be readily incorporated, provided analytical models can be developed.

Generic Failure Rate vs. Age Curves

It should be emphasized that the Monte Carlo simulation model is strictly applicable to ***110 SDR 11 PE pipes only***, since this is the pipe size that the inherent defect size data applied. Furthermore, it should be noted that the fracture properties used in the model apply strictly to the particular samples used in this study and ***should not be assumed to apply to all PE materials of that designation***. Although different PE materials can attain the same designation (in terms of Minimum Required Strength category or Hydrostatic Design Basis), significantly different mechanical properties and SCG resistances can be observed. Further work is required to quantify the variation in material and fracture properties within different PE materials grades.

Extended Simulation Times for PE100 (AUS) and PE3408 (US)

The relatively low failure rates in PE100 (AUS) and PE3408 (US) pipes are attributed to extended simulated failure times, which in turn, are attributed to the increased long term craze strength of these materials. However, the crack initiation/net section collapse failure model developed in this study does not account for chemical stabiliser loss and the onset of brittle failure associated with this loss, should it occur. Further work should be conducted to identify the "chemical knee" and the onset of stage III brittle failures in these materials at ambient temperatures.

CHAPTER 1
PRODUCTION OF POLYETHYLENE PRESSURE PIPES

INTRODUCTION

The unique characteristics of plastic pipes such as high strength to weight ratio, chemical and corrosion resistance, ease of installation and flexibility have been sited as advantages over traditional metallic and cement-based materials. Polyethylene (PE) pipe, the second most used polymeric material in pipe and conduits, was first introduced in the United States gas transportation industry in the 1950s. Since then, the demand for PE has grown and it is now used for water distribution, irrigation and storm water pipes.

Basically chemically inert, PE overcomes the major problem associated with all metallic pipes over the ages: corrosion. With whole industries founded on preventing corrosion in pipelines, PE can provide a major cost-cutting advantage for the water and gas transportation industry.

A thermoplastic polymer, PE has a relatively low density, melting temperature, stiffness and strength. A comparison of published physical properties for PE and metallic pipeline materials is shown in Table 1.1 (Bowman, 1995).

The high strength to weight ratio implied by these properties provides PE with some important advantages for pipe applications. For example, its lower density reduces pipeline weight, facilitating handling and transportation. A lower tensile (Young's) modulus increases flexibility, allowing large deformations to be tolerated with no drop in load bearing capacity. This is best illustrated by the recent service history and survival of PE pipelines under sever ground loading, notably the Kobe earthquake of 1995. High flexibility also allows long lengths of PE pipe (up to 7.1 inches, or 180mm in diameter) to be coiled immediately after production, which eliminates a considerable number of joints from a finished pipeline. Perhaps the most important economic advantage of this increased flexibility is illustrated by the development of cost-saving installation procedures exclusive to PE. For example, directionally controlled 'moles' are now used which open the earth whilst pulling the pipe along behind, in an almost continuous laying procedure. Similar technology is used for lining corroded metallic pipelines in renovation projects.

THE DEVELOPMENT OF POLYETHYLENE FOR PRESSURE PIPE APPLICATIONS

Polyethylene pipe materials have often been distinguished by density. Typical materials used for pipe applications are:

- Low density LDPE (density < 930 kg/m^3) (58.06 lb/ft^3)
- Medium density MDPE (density 930 < 940 kg/m^3) (58.06 < 58.68 lb/ft^3)
- High density HDPE (density > 940 kg/m^3) (58.68 lb/ft^3)

Polyethylene in raw material form is produced by the polymerization of ethylene obtained from petroleum feedstock. Ethylene gas is heated under pressure and single units (monomers) of ethylene (-[CH_2]-) are chemically linked to each other in a catalyst-initiated reaction. A polymer chain molecule is produced with a length or molecular weight dependent on the number of

Table 1.1
Physical properties of pipe grade PE and selected metals (adapted from Bowman, 1995)

	Polyethylene	Metallics
Density, lb/ft^3 (kg/m^3)	57 to 60 (920 to 955)	162 to 556 (2,600 to 8,900)
Melt Temperature,°F (°C)	<266 (<130)	1,112 to 2,732 (600 to 1,500)
Young's Modulus, psi (GPa)	43,511 to 145,038 (0.3 to 1.0)	10.15×10^6 to 30.46×10^6 (70 to 210)
Yield Stress, psi (MPa)	2,176 to 3,626 (15 to 25)	7,252 to 72,519 (50 to 500)

ethylene units polymerized. PE materials are composed of a distribution of molecular weights, appropriate to their end use.

For new pipe-grade PE materials, the repeated CH_2 chain structure is usually modified, and it is generally accepted that these materials are actually *branched copolymers*, with foreign monomers grafted into their main ethylene chains during polymerization. For example, propylene, butene or hexene is added, forming side branches along the main ethylene chain. The length and position of these branches influence some extremely important properties of a pipe-grade material.

The "modality" of a Polyethylene material refers to the form of its molecular weight distribution curve (a plot of polymer weight fraction as a function of molecular weight). Conventional, *unimodal* PE materials are produced by a single polymerization reactor and exhibit a single maximum point as shown in Figure 1.1. Unimodal pipe grade PE materials typically exhibit a comonomer side branch content that is segregated to the low molecular weight side of the molecular weight distribution (Figure 1.1). In contrast, new, third generation *bimodal* pipe grade PE materials are produced in a serial two-stage polymerization process where separate, independent reactors are used to produce low and high molecular weight distributions. A low molecular weight homopolymer (with no comonomer content) is produced in the first polymerization reactor and a high molecular weight copolymer (with comonomer side branching) is produced in the second reactor. The resulting PE material has a molecular weight/comonomer distribution which is superposition of these separate distributions. In general, the molecular weight distribution is relatively wide and may, in some cases exhibit two separate peaks (Berthold et al. 2001). As shown in Figure 1.1, the side branch content of bimodal materials is biased towards the high molecular weight side of the distribution. This change from unimodal to bimodal PE materials results in improved resistance to slow crack growth and rapid crack propagation (Berthold et al. 2001).

PE Development in Europe

All plastic materials used for the manufacture of pipeline systems in Europe are classified for long term performance under hydrostatic pressure in accordance with the international standard for "Thermoplastics materials for pipe and fittings for pressure applications—Classification and designation—Overall service (design) coefficients" ISO 12162 (International Organization for Standardization, 1995). This classification system is based on the predicted Minimum Required Strength (MRS), which is the estimated 97.5% lower prediction limit of the average long term strength extrapolated to 50 years. The MRS is a long term strength category and to enter an MRS category, a material must exhibit a specified minimum value of long term strength in accordance with "Plastics Pipes and Ducting Systems—Determination of the Long-Term

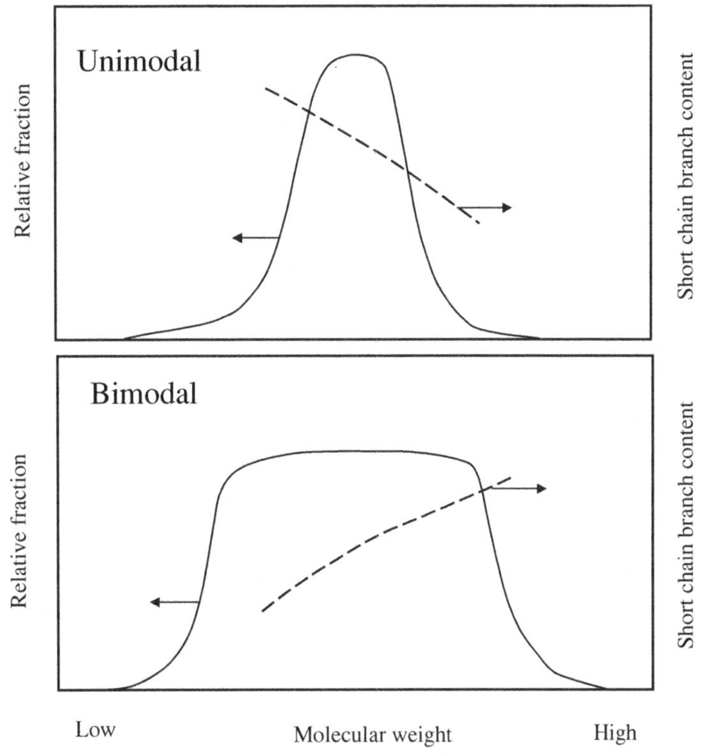

Figure 1.1 Schematic illustration of molecular weight and comonomer distributions in conventional and new generation pipe-grade PE materials

Hydrostatic Strength of thermoplastics Materials in Pipe Form by Extrapolation" ISO 9080 (ISO, 2003). This required long term hydrostatic strength (LTHS) is the sustained hydraulic stress that, on average, would result in failure after 50 years.

In Europe, the first PE water transportation systems were installed in the 1950s using "first-generation" low density (LDPE) and high density (HDPE) polyethylene. LDPE was initially used extensively for small diameter water service pipes, but was replaced in the early 1980s. First generation LDPE pipes were PE32 classified with an MRS of 3.2 MPa (464.1 psi) and therefore required a relatively large wall thickness for a given pressure. The higher hoop strength and stiffness of early HDPE grades enabled large diameter pipelines to be designed with reduced wall thicknesses, but both materials exhibited small "slit" failures which are the result of a brittle-like process of crack initiation and slow crack growth.

It is important to note that classifying a PE material based solely on its projected long term hydrostatic strength may not allow field performance under service conditions to be forecast. Additional information is required to quantify the resistance of a particular material to crack initiation and slow crack growth before a more reliable projection of likely field performance can be made. The main advantage of these early PE80 HDPE materials is their low cost and high short-term strength.

Research into PE was continuous, and by the late 1970s, a second generation medium density polyethylene (MDPE) was produced which maintained a relatively high stiffness without displaying brittle-like characteristics. Second generation MDPE materials were developed in response to a requirement of the gas distribution industry for PE materials with improved resistance to slow crack growth. Since these materials have gained PE80 classification, it was important to distinguish between PE80 MDPE and PE80 HDPE. As stated above, these materials display significantly different slow crack growth resistances.

In the early 1990s, bimodal HDPE materials (Figure 1.1) were developed in Europe with higher hoop strength giving rise to the PE100 classification (MRS = 10.0 MPa (1,450.4 psi)). These materials produced stronger pipes used for higher pressure operation in gas and water distribution systems. Compared to PE80 materials, the pipe wall thickness can be reduced, increasing the cost effectiveness of polyethylene against ductile iron at larger diameters. However, the apparently 25% greater hydrostatic strength of PE100 materials compared to PE80 materials should be interpreted in the context of the ISO 9080 long term strength categorization (ISO, 2003). As stated above, this system assigns materials to a strength category and while some PE80 materials may exhibit long-term strengths significantly in excess of 8.0 MPa, some PE100 materials may just meet the requirements of the PE100 category. Therefore, the perceived 25% increase in long term strength is not applicable in all cases.

Initially Solvay, a Belgian supplier was the leading producer of third generation PE100 materials but other manufacturers now offer these materials commercially. Founded on February 24, 1999, the "PE100+ Association" is an industry organization of several PE material manufacturers (Innovene, Prime Polymer, Sabic, Total Petrochemicals, Borealis A/S and Basell) which monitor the most critical properties and requirements of PE100 materials such as creep rupture strength, stress crack resistance and resistance to rapid crack propagation. The PE100+ Association regularly post a "Positive List of Materials" on the Internet at www.pe100plus.net and various independent laboratories repeat material tests every seven months.

PE Development in Australia

The first Australian PE material was introduced for rural applications in the 1950s and designated "Type 50" equivalent to an MRS of 6.3 MPa (725.2 psi). These materials were replaced by first generation HDPE PE80C grades as water authorities began using PE for gravity and pressure sewer projects. PE became more widely used in the mid 1970s when general approval for plastics usage in the water industry was obtained. Initially, pipes were mainly restricted to small diameter plumbing applications, but the introduction of electrofusion fittings facilitated the use of PE in larger diameters. Although PE had been widely used by the Australian gas industry since the early 1970s, its widespread use for water transportation did not occur until the 1980s, with the introduction of second generation MDPE materials attaining PE80B rating in accordance with AS 4131(Int) (Standards Australia/Standards New Zealand, 2003a). In the 1990s, PE100 materials conforming to AS 4131 (Standards Australia/Standards New Zealand, 2003a) were introduced into Australian water utilities. These materials are bimodal, third generation PE grades, such as those produced in Europe.

As outlined in the current Australian standard for Polyethylene (PE) compounds for *pressure* pipes and fittings (AS 4130) (Standards Australia/Standards New Zealand, 2003b), PE materials are classified as PE80 and PE100 according to ISO 12162 (ISO, 1995). PE80 materials are then further subdivided into PE80B and PE80C type materials, based on their hydrostatic strengths at elevated temperature (Table 1.2).

PE80B and PE80C materials are also distinguished in terns of their resistance to slow crack growth when tested in accordance with the international standard for "Polyolefin pipes for the conveyance of fluids—Determination of resistance to crack propagation—Test methods for slow

Table 1.2
PE pipe-grade materials in Australia and their minimum lower limits of predicted hydrostatic strength for design life of 50 years

Temp, °C (°F)	Lower limit of predicted hydrostatic strength, MPa (psi)		
	PE80B	PE80C	PE100
20	8.0 (1,160.5)	8.0 (1,160.5) (50 yrs)	10.0 (1,450.6)
25	7.5 (1,088.0)	7.2 (1,044.4) (50 yrs)	9.4 (1,363.6)
30	7.0 (1,015.4)	6.5 (942.9) (28 yrs)	8.7 (1,262.0)
35	6.4 (928.4)	5.8 (841.3) (16 yrs)	8.0 (1,160.5)
40	6.0 (870.4)	5.0 (725.3) (9 yrs)	7.3 (1,058.9)

crack growth on notched pipes (notch test)" ISO 13479 (ISO, 1997). Minimum allowable test times for PE80 C materials are lower than those for PE80 B materials.

PE80B and PE80C materials are also distinguished in terms of their resistance to slow crack growth when tested in accordance with the international standard for "Polyolefin pipes for the conveyance of fluids—Determination of resistance to crack propagation—Test methods for slow crack growth on notched pipes (notch test)" ISO 13479 (ISO, 1997). Minimum allowable test times for PE80 C materials are lower than those for PE80 B materials.

PE Development in the USA

In comparison to European and Australian performance classification systems for PE pipe material, American classification systems are more complex. This greater complexity results from the following objectives which were originally set in the 1960s and are still begin followed (Mruk, 2006).

- The identification of the nature of a piping material by the use of standard abbreviations (i.e., "PE")
- The identification of properties, other than a material's projected long term strength that are believed to be important elements in the performance requirements for different kinds of pressure pipe applications. This property identification is achieved by appending two or three digits following the material abbreviation
- The identification of a material's maximum recommended hydrostatic design stress (HDS) for water at a temperature of 73.4°F. This done by the addition of two last digits which code the HDS in units of 100 psi.

An important difference between the American and the ISO material designation systems is that in the former, the maximum recommended hydrostatic design stress is identified. This convention is in line with a common engineering practice whereby a recommended maximum design stress for a material is established for a given application and service conditions. A reason for this practice is that the numerical value of the "safety factor" which is used to reduce a certain laboratory established strength into a design stress often conveys a greater sense of security than can actually be justified. This is because the laboratory determination of a material's strength often fails to consider all the stresses and the different conditions that a material will encounter in actual service. In respect to the second bullet point, a gradual evolution has occurred in the identification of those PE properties that have been identified as significant to quality of field performance (Mruk, 2006).

As outlined in American Water Works Association standard for "PE pressure pipe and fittings for water distribution and transmission" ANSI/AWWA C-906 (American Water Works Association, 1999), pipe and fittings materials are assigned one of three standard PE code designations:

- PE2406
- PE3406
- PE3408

In turn, these AWWA standard code designations refer to a PE material minimum cell classification, as prescribed by the ASTM standard specification for "Polyethylene plastics pipe and fittings materials" ASTM D 3350-02a (ASTM International, 2002). This cell classification consists of the letters "PE" followed by six numbers and a single letter. As described in ASTM D 3350 (ASTM International, 2002), the numbers are codes for minimum allowable values for six material/pipe properties and single letter corresponds to the color and UV stabilizer of the material. The properties included in the cell classification are:

(1st number)—Density
(2nd number)—Melt Index
(3rd number)—Flexural Modulus
(4th number)—Tensile Strength
(5th number)—Slow Crack Growth resistance
(6th number)—Hydrostatic Strength classification

In accordance with ASTM D 3350 (ASTM International, 2002), Slow Crack Growth resistance is specified based on results from either test method ASTM D 1693-01 "Environmental stress cracking of ethylene plastics" (ASTM International, 2001a) or test method ASTM F 1473-01 "Notch tensile test to measure resistance to slow crack growth of Polyethylene pipes and resins" (ASTM International, 2001b). ASTM D 1693 is referenced in cases of lower slow crack growth requirements while F 1473 is referenced for higher levels of resistance such those available in third generation materials.

Hydrostatic strength is given by the Hydrostatic Design Basis (HDB) as determined by the standard test for "Obtaining hydrostatic design basis for thermoplastic pipe materials or pressure design basis for thermoplastic pipe products" ASTM D 2837-02 (ASTM International, 2003). In accordance with method D 2387, the HDB is the strength category in which a material is placed, based on its long term strength (LTHS). The LTHS is determined by the extrapolation of the stress regression line to the 100,000 hour intercept. The HDB is therefore a hydrostatic design basis which, when multiplied by a design factor, yields the HDS. Because in method D2837 the extrapolation is carried to the 100,000 hour intercept some have falsely concluded that the design life under this method is only 100,000 hours (11.4 years). When developing this method the intent was to consider the 100,000 hour intercept only as a design basis which needed to be reduced by an appropriate design factor so as to yield an HDS that provides a service life that is very much longer than just 100,000 hours. The long term strength basis and the design factor are interdependent. Furthermore, it is not believed that a method for the forecasting of a material's hydrostatic

Table 1.3
PE pipe-grade materials in the USA and their minimum HDB requirements

AWWA C909 Standard PE code designation	ASTM D 3350-02a cell classification	ASTM D 2837-02 Hydrostatic Design Basis, psi (MPa)
PE2406	PE213333C, D or E	1,250 (8.62)
PE3406	PE324433C, D or E	1,250 (8.62)
PE3408	PE334434C, D or E	1,600 (11.03)

strength based on just one kind of laboratory test should be the sole instrument for the predicting of the design life of a plastics pipe.

Table 1.3 lists the PE material classifications according to AWWA C909 and the corresponding material cell classifications in ASTM D 3350-02a (ASTM International, 2002). The Hydrostatic Design Basis is also given for each material type. It should be noted that ASTM D 3350 has recently been revised to give specific recognition to those newer PE piping materials that offer very high resistance to slow crack growth and, as such, allow the safe use of a higher design stress. Work has been initiated at standards writing groups to update PE pipe standards so as to recognize these newer material classifications (Mruk, 2006).

As shown in Table 1.3, PE2406 and PE3406 materials have the same minimum requirements for Hydrostatic Design Basis (HDB) whereas PE3408 exhibits a higher Hydrostatic Design Basis of 1600 psi (11.03 MPa). Further examination of ASTM D 3350-02a (ASTM International, 2002) also indicates that PE2406 corresponds to MDPE materials with relatively low values of density, melt index, tensile strength and flexural modulus compared to the High Density PE3406 and PE3408 materials. Apart from an increased HDB, requirements for PE3408 grades are distinguished from PE3406 by a lower Melt Index which may indicate an increased molecular weight in PE3408 materials.

The development of PE materials in North America was driven by the gas industry and the need to reduce the likelihood of brittle fracture. HDPE grades were the first PE materials to be developed in North America in the mid to late 1950s. Although these materials exhibited the desired properties for pipe applications (adequate strength, stiffness etc.), they were also susceptible to brittle-like fracture. Hydrostatic strength testing at elevated temperatures suggested that these materials would exhibit a downturn in stress rupture behavior before the desired design life at ambient temperature. Such a downturn is known to result from a relatively low resistance to slow crack growth, and helped explain the relatively poor field performance of some early HDPE grades labeled PE3306. This led to the abandonment of PE3306 and the establishment of the PE3406 category. However, it should also be noted that as early as the 1950s a high density PE pipe was offered to the gas distribution market which was made from a very high molecular weight material. This material offered outstanding resistance to slow crack growth and in elevated temperature tests, (up to 90°C), there was no indication of a downturn in testing times that covered about 20,000 hours. However, because of its very high molecular weight this pipe could not be easily joined by heat fusion and connections were made using compression type fittings. In the early 1960s, when a heat fusion system was first introduced, this higher molecular weigh PE pipe was used less frequently. However, the significant point is that almost 50-years later it has been reported by early users of this pipe that it still giving very reliable service. Again, this suggests the importance of rating a PE pipe grade not just by the estimated long term hydrostatic strength but also by its ability to resist failure by slow crack growth.

In the early to mid 1960s, PE materials were produced with improved resistance to brittle-like failure. For example, Du Pont produced an MDPE copolymer with foreign side chains introduced to the main ethylene chains (as in Figure 1.1), which improved brittle crack resistance. In the late 1970s PE materials were produced with an increased hydrostatic design basis and a change from AWWA classification PE3406 to PE3408 occurred. Table 1.3 indicates that this change would have resulted in an increase in Hydrostatic Design Basis from 1,250 psi (8.62 MPa) to 1,600 psi (11.03 MPa). In the late 1980s to early 1990s third generation bimodal PE materials were introduced to the North American polyethylene piping industry to compete with the existing unimodal materials. In North America, pressure pipe made from the new third generation materials continued to be pressure rated based on the same protocol as used for the older materials. As stated above, a policy has been adopted that allows the use of a higher design factor multiplier in the case of any PE materials that is distinguished by very high resistance to Slow crack growth. This covers not only the so called PE100+ materials but also other PE materials that exhibit superior fracture toughness and ductility.

PE PIPE EXTRUSION

PE is converted from raw material to pipe by extrusion. In this process, polymer pellets are released into a barrel zone containing the extruder screw, where they are converted into a homogeneous melt. Rotation of the screw produces heat by mechanical working of the raw material which, when combined with external heating of the barrel increases its temperature to around 392°F (200°C). The pellets that are fed into an extruder consist not only of polymer (resin) but of PE resin that has been pre-compounded with the proper amounts of ultra-violet screens or stabilizers, anti-oxidants, extrusion aids and colorants. In some cases the colorant consists of a separate colorant/resin masterbatch which, along with the un-colored PE composition, is carefully metered into the extruder hopper so as to produce a well compounded mixture having the desired coloration (e.g., yellow for gas distribution applications).

While generating heat, screw rotation also transports the molten polymer along the extruder barrel and forces it through an annular die with the initial shape of the pipe cross-section. Pull-rolls, located further down the extrusion line, apply tension to counteract visco-elastic expansion of the extrudate before it is passed onto a sizing and cooling stage. The principal role of the pull-rolls is to develop a force that pulls the pipe at a controlled rate through each of the stations after it leaves the die, including diameter sizing, cooling and pipe cut-off. Uniform pulling is an essential factor in the close control of wall thickness.

In most cases, the extruded pipe is cooled by water which is sprayed onto its outer surface in a series of tanks located downstream of the die exit. During this single-surface cooling process a gradient of cooling rates is set up through the pipe wall thickness. An outer "skin" of polymer solidifies first, followed by the inside surface, cooling at a much lower rate. Although this method is used by many pipe manufacturers, an alternative exists, dating back to 1970. Introduced by DuPont in the US, "dual-surface cooling" was developed to combat sagging of molten pipe, extruded from PE materials with a low melt viscosity. For these materials, gravity-induced flow (or pipe sag) occurred at the inside surface of the pipe, which remained molten for long periods during single-surface cooling. This could often result in non-uniform pipe wall thicknesses. By cooling rapidly on both external and internal surfaces, dual-surface cooling allows a solidified shell to be formed which sets the cross section while slowly cooled material at the middle of the pipe wall solidifies later. Simultaneous cooling of the pipe outer and inner surfaces is now seldom applied since modern PE piping materials exhibit sufficient melt-strength to avoid the problem of "pipe-sag."

Upon cooling from the melt, PE adopts a semi-crystalline form comprised of an ordered crystalline phase, in which molecular segments are closely packed in three-dimensional arrays. Molecular chain axes are oriented in one particular direction and groups of these arrays form crystalline lamellae arranged spatially throughout the material. However, because a polymer melt contains an enormous number of chain entanglements, the organized packing procedure required to form a 100% crystalline solid cannot be achieved. PE is typically comprised of a 40% to 70% crystalline mass fraction existing within a surrounding amorphous phase of disordered chain segments and molecular entanglements. While, (as discussed below) cooling history has some effect, the crystalline fraction is mainly controlled by the architecture of the polymer molecule. Fewer and shorter side branches allow molecules to pack closer together, which leads to an increased crystalline phase that result in a higher polymer density and higher stiffness and higher short-term strength. Most PE pipe resins exhibit densities on the higher side, as compared to lower density PE's that are used for other applications.

Post-Extrusion Cooling of PE Pipes

The cooling rate of a polymer melt during extrusion influences the degree of crystallinity attained in a finished pipe product. In general, a slower rate of cooling increases the degree of crystallinity and a rapid cooling rate reduces crystallinity.

For single-surface cooled pipe the lowest crystallinity is observed at the outer surface of the pipe and the highest crystallinity at the inner surface. For dual-surface cooled pipe the lowest crystallinity is attained at the inner *and* outer surface of the pipe with a maximum at the mid-point of the pipe wall. In general, an increase in crystallinity produces increases in Young's modulus and Yield strength (Davis et al. 1998). However, the existence of a crystallinity profile across the wall of a PE pipe has not been linked with any adverse effect on long-term properties, such as hydrostatic strength or resistance to slow crack growth.

In common with most polymers, PE has a low thermal conductivity, and any non-uniform pipe cooling process produces substantially different cooling rates through the pipe wall. Consequently, local density changes and accompanying thermal contractions of molten material vary at different positions through the pipe wall. During single-surface cooling, material at the pipe outer surface solidifies quickly, preventing contraction of the slower cooling material at the pipe inner surface. This inevitably leads to the presence of residual stresses, through the action of "frozen-in" strains and the modulus of the cooled material. The outer layers of the pipe are in compression, balanced by the inner layers in tension. Dual-surface cooling results in rapid solidification at the external and internal surfaces of the pipe and therefore, a symmetrical residual stress profile through the wall thickness of the pipe. The outer layers of the pipe are in compression and are balanced by a central region of residual tension where material solidifies slowly.

As with other thermoplastic polymers, this residual stress at the pipe inner surface is thought to influence fracture failure modes for PE pipes in service. In some cases, the magnitude of a residual stress resulting from a manufacturing process can sufficiently add to the overall stress so as to quicken the development and growth of cracks that can eventually lead to fracture. However, this quickening effect is not consistent among all kinds of PE materials. In the case of the early PE's that exhibited low slow crack growth resistance, the effect of residual stress was believed to be significant. In the case of modern PE materials which show increased resistance to slow crack growth, tests show that they can readily absorb localized stress intensifications that are caused by residual stresses without any adverse effect on their long-term properties. It should also be noted that many standards, such as AWWA C906, include a requirement which limits the extent of residual stress that can be present on the inside surface of PE pipe.

JOINTS AND FITTINGS IN PE PIPES

In addition to mechanical flanged or compression sealing joints, PE pipeline systems can be joined using butt-fusion or electrofusion welding. Both these techniques are described in detail below.

Butt Fusion Jointing

In the butt-fusion jointing process, the pipes to be joined are mounted in the clamps of the butt fusion equipment and checked for initial alignment. The pipe ends are then planed to ensure that they are flat and square. The jointing sequence begins with a flat hot plate (at a preset temperature), is positioned between the two pipe ends. The pipes are then pushed towards each other until the pipe ends come into contact with the hot plate and the pressure is increased to give good thermal contact. The pipe ends melt and the interface pressure forces the molten material outwards to form "weld beads" at the outside and inside pipe surfaces. This is often referred to as the "bead-up" stage of the process. At the end of this stage, the pressure is reduced to only just maintain contact between the pipe and the hot plate. This allows the depth of molten material along the pipe length to increase without increasing the size of the weld beads. This stage is often referred to as the "heat soak." The ends of the pipe are then pulled away from the hot plate and the hot plate is removed. The two molten pipe ends are pushed together at same pressure as that used during the initial bead-up stage. This produces further growth of the weld bead causing it to curve over towards the pipe external surface. This stage is often referred to as the "bead roll over" stage. The pressure is maintained until the weld is fully cooled. Butt fusion welding of thermoplastic pipes can be carried out on a wide range of pipe sizes, typically between 63mm (2.5 inch) and 1,600mm (63 inches) outside diameter. Manual, automatic and semi-automatic butt fusion welding machines are available commercially from a number of suppliers.

In a recent study of the quality of butt-fusion welds in large diameter thick-walled PE pipes, Hill, Wilson, and Headford (2001) state that three external factors are crucial to the formation of a good butt-fusion welded joint.

- *Absence of contamination*—The ends of the pipe prior to contact must be free from any contamination, such as airborne dust particles, plant debris, machine oil or lubricants. During welding PE in hot, dry weather (low humidity), the material has a tendency to generate significant levels of surface static during surface planing, which can then attract dust (Hill, Wilson, and Headford, 2001). Local airborne dust levels can also increase suddenly and significantly when welding machines are not operated smoothly and progressively (Hill, Wilson, and Headford, 2001).
- *Sufficient heat input*—The thermal capacity of the pipe ends must be sufficient to ensure that enough heat energy is available for good fusion to take place. Since the

specific heat capacity of the pipe material is constant, this means that sufficient mass of molten polymer must be created at the required temperature. Large, thick-walled pipes require increased heat energy, which means that heat soak times are extended.

- *Good final morphology*—Since PE is a semi-crystalline polymer, the morphology (crystalline structure) of the weld zone area is critical to performance. The pipe morphology can be completely changed in the vicinity of a weld, which is particularly important for thick pipe wall sections were sufficient depth of molten polymer may not have been achieved during heat soak stage.

It is important to note that these external factors are recognized and addressed by standards (i.e., ASTM D2657—Standard Practice for heat Fusion Joining of Polyolefin Pipe and fittings) and industry organizations. For example, the Plastics Pipe Institute (PPI) provides technical guidance on joining of PE pipes (PPI TR 33 Generic Butt Fusion Joining Procedure of Polyethylene Pipe, http://plasticpipe.org/pdf/pubs/reports/TR-33-05.pdf). It is general practice—and a mandated requirement in the case of gas distribution applications—that the contractors making butt-fusion joints must first have been evaluated, and qualified, for their proficiency in the recommended procedure (Mruk, 2006).

Several important performance properties of PE pipes (such as slow crack growth resistance and impact crack initiation resistance) depend on the high molecular weight of the material. It follows that in order for a butt-fusion joint to exhibit the required performance properties, PE molecules from both pipe ends must mix prior to weld cooling. However, the pipe material at the joint interface will not usually have sufficient time during the fusion process to become perfectly mixed (Hill, Wilson, and Headford, 2001). The material at the weld interface will therefore behave as though it was a PE material of apparently lower molecular weight. This is reflected in comparisons of impact strengths for welded joints in PE80 and PE100 materials grades in the UK (Hill, Wilson, and Headford, 2001). In both cases, the measured impact strength of welded joints was below that of the parent pipe. However, the greater deficiency between joint and pipe impact strength was observed in PE100 grades, which exhibit a higher molecular weight than PE80. This was attributed to the reduced mixing (and hence lower apparent molecular weight) created at the joint interface in these materials (Hill, Wilson, and Headford, 2001).

However, it should also be noted that impact testing subjects a fusion joint to a different stress scenario than that experienced under normal operating loads. While a reduction in impact strength may be observed, this should not be automatically interpreted as a corresponding reduction in long term strength. In fact indications are that properly made butt-fusions perform well in these tests. Parmar and Bowman (1989) state that "properly made joints have strengths similar to pipe, when tested at elevated temperatures under constant or fluctuating pressures, or by the simple tensile test."

Electrofusion Jointing

The electrofusion (EF) jointing process involves the use of a sleeve fitting (coupling), which contains resistance wires carrying an electrical current. The two pipe ends to be joined slide into the coupling, which contains an internal stop to prevent the ends from meeting. When the resistance wires are heated by the passage of a controlled electrical current, the surrounding PE material melts and on subsequent cooling, forms a fusion joint. Fusion indicators are commonly designed into the sleeve fitting, and protrude when sufficient melt pressure has been generated through expansion of the pipe onto the coupling inner surface.

It is widely acknowledged that, in order to establish a consistent and structurally sound joint, it is necessary to follow a strict preparation procedure. If the appropriate procedures are followed, contamination and disturbance effects that might inhibit the fusion mechanism will be minimized. For successful joining of pipes, three important pipe preparation stages must be followed. Firstly the pipe ends must have finished squared ends to ensure that the central cold zones function to contain the melt. Secondly, the pipe surfaces to be joined must be properly scraped to reveal uncontaminated material. With the EF joining process, there is little or no relative movement between the pipe and the coupler. Therefore, any contamination on the pipe surface is retained at the joint interface, which can significantly reduce the strength of the joint. Finally, the pipe and fitting should be clamped during welding to eliminate relative movement. This ensures that the molten polymer is contained at the fusion interface, allowing the development of a strong joint. The American standard ASTM F1290, "Standard Practice for Electrofusion Joining Polyolefin Pipe and Fittings" recognizes the importance of proper practice and detailed electrofusion guides are issued by manufacturers of electrofusion fittings. To insure that proper practice is being followed all fitting manufacturers offer a control box which regulates each of the essential electrofusion steps.

In a study of EF coupling fittings under hydrostatic pressure, Allcard and Beech (2001) assessed the effect of fitting material on long term failure. EF fittings manufactured from PE80 and PE100 materials were welded to PE100 pipes and pressure tested at 80°C. Test results indicated that no significant reduction in the Slow Crack Growth resistance of fittings occurred even though parameters such as injection molding temperature and wire temperature were varied significantly.

INSTALLATION OF PE PIPELINES

As discussed previously, the high flexibility of PE pipelines facilitates relatively easy installation and has led to the development of novel, trenchless installation technologies. The Plastics Pipe Institute (PPI) in the US have published a free downloadable "Handbook of Polyethylene Pipe" which in addition to a providing a description of Polyethylene materials, pipe manufacturing and design codes, also describes methods for pipe installation. Like all polymeric materials, Polyethylene is sensitive to changes in temperature and will attempt to expand as temperature increases and contract as temperature decreases. The PPI handbook provides a brief description of thermal expansion and contraction effects in PE and the incorporation of thrust restraints in the design of PE pipelines (PPI, 2007). The PPI Handbook also gives a detailed description of the design of flexible pipelines, including trench design and construction, embedment and compaction techniques and the behavior of different trenching materials (PPI, 2007). The PPI handbook also addresses marine applications with a detailed description of underwater pipe installation and anchoring methods.

WATER QUALITY IN PE PIPES

With increased market participation, the integrity of water transported through polyethylene pipes has been examined. Potable water delivered to the consumer must be fit for human consumption. Hence, materials used in the distribution system should not be detrimental to human health, nor impart an objectionable taste or smell. To this end, it is necessary to evaluate the material interaction with water, and in particular those elements in the material formulation that could have adverse effects on water quality.

PE pipe is manufactured by extrusion of polyethylene resin, generally at temperatures ranging between 190 to 250°C, followed by cooling in a water bath or air (Janson 1995). The pipe resin formulation requires the incorporation of a range of additives to ensure long-term stability:

- UV stabilizers or screeners for weathering resistance
- Antioxidants (thermal stabilizers) for heat resistance during processing and aging
- Flame retardants (not used in North America)
- Pigments
- Lubricants and extrusion aids, etc.

In the US, concern about potential adverse effects on water quality received attention from the first time that plastics pipe was commercialized. In 1951, a program was initiated with the National Sanitation Foundation (NSF) to identify any effects and make appropriate recommendations. In 1955, the NSF issued a resultant report titled "A Study of Plastics Pipe for Potable Water Supplies." Based on this report, materials intended for water applications would only be granted an NSF seal if toxicity and other tests determined that it was safe for the transport of potable water. In 1965, the NSF published the first issue of NSF standard 14 which included updated public health and safety requirements for plastics piping. Compliance to this standard was a requirement by States and other governing authorities.

In 1985, the US Environmental Protection Agency (USEPA) entered into a cooperative agreement with NSF to develop voluntary third-party consensus standards and a certification program for all direct and indirect drinking water additives. The American Water Works Association was member of the resultant consortium. This effort produced ANSI/NSF 61, (Drinking Water Systems Components—Health Effects. Annex A, "Toxicology Review and Evaluation Procedures") which indicates a maximum allowable level (MAL) of contaminants for substances not regulated by a USPA final maximum contaminant level (MCL). The MAL for an unspecified list of "unregulated contaminants" are based on toxicity testing guidelines (noncarcinogens) and risk characterization methodology (carcinogens). All PE pipe compositions offered in North America are designed to comply with the requirements of ANSI/NSF 61.

While there is limited research on the effect of PE pipes on water quality, extensive research on the effect of PE packaging materials on food and bottled water quality is drawn upon in the selection of additives that are used in PE pipe compositions (Mruk, 2006). Among the components used in the formulation of PE pipes, additives such as antioxidants, lubricants, stabilizers and their by-products formed during processing are the main components found to migrate into water. A review of literature from the EPA, FDA, WHO, EC water guidelines and environmental agencies (For example, Greenpeace and Friends of the Earth) did not indicate that the additives currently used in PE pipes constitute a health hazard.

Odor and off-taste incidents appear to be the main cause of water quality complaints associated with PE pipe, even if the contaminant concentrations are safe for human consumption. There is no evidence to suggest that PE promotes growth of microorganisms or biofilms compared to other types of pipe.

CONCLUDING REMARKS

Due to a high strength-to-weight ratio and increased flexibility compared to other materials, the use of PE for pressure pipe applications is increasing worldwide. The current generation materials have attained a PE100 rating corresponding to a Minimum Required Strength (MRS) category of 10 MPa (1,450 psi) at 20°C. To be qualified for a certain MRS category, a material must exhibit a long term hydrostatic strength (LTHS) in accordance with method ISO 9080 (ISO, 2003) that is not lower than the MRS category into which the material is placed. ISO 9080 defines an LTHS as the lower predicted value of hydrostatic stress at which failure is predicted to occur, on **average**, in 50-years. Therefore, to ensure failure free operation **prior** to 50-years, the sustained hydrostatic stress will have to be somewhat lower than the predicted LTHS. In addition to MRS rating, supplementary information is also required fully characterize the expected field performance of a PE pipe material. For example, new PE materials maintain high levels in important properties such as creep rupture strength, stress crack resistance and rapid crack propagation resistance. In the US, standard specifications are updated for new materials to include requirements such as a minimum resistance to slow crack growth, which in turn, ensure that the required service life is achieved.

In addition to mechanical jointing, PE pipelines can be joined by butt-fusion welding or electrofusion coupling. In butt-fusion welding, the ends of the pipe to be joined are planed and prepared then brought in to contact with a heating plate under pressure. After a pre-determined heat soak period, the plate is removed and the ends are joined together and allowed to cool. When the butt fusion process is conducted properly, it results in joint strengths equivalent to those of the pipes being joined. Electrofusion (EF) jointing involves the use of an outer sleeve fitting, or coupling, which contains resistance wires carrying an electrical current. When the resistance wires are heated by the passage of a controlled electrical current, the surrounding PE material melts and on subsequent cooling, forms a fusion joint. Assuming good basic design and installation, fittings manufactured from PE80 and PE100 grades (and joined to PE100 pipe) exceed minimum requirements tested at elevated temperatures.

Among the components used in PE pipe manufacture, additives such as antioxidants, lubricants and stabilizers are the main components found to migrate into water. However, these do not constitute a health hazard and there is no evidence to suggest that PE promotes growth of microorganisms or biofilms compared to other types of pipe.

CHAPTER 2
FIELD PERFORMANCE OF POLYETHYLENE WATER PIPES

INTRODUCTION

For any engineering application, the most important measure of effectiveness is the actual performance that is achieved in the field. To this end, this chapter presents the responses to an industry survey that requested data concerning the rates of recorded failures in PE pipe systems and the modes of failures that occur.

However, it is essential to note that the quality of this information is variable across the water industry. A further complication is that different generations of PE materials have been available over the ongoing timeline of PE pipe installation. As discussed in chapter 5, earlier material grades can be expected to exhibit relatively poor slow crack growth resistance compared to the currently available PE material grades. Relatively long service lifetimes and low failure rates can be expected from these current grades of PE. However, the necessary information to distinguish these material grades does not exist within the water industry. For example, although a failed pipe can be identified as "PE80" in a particular water authority database, the additional information to determine the particular generation of the failed pipe is unavailable.

It should also be noted that the failure of a pipe in service is influenced not only by the inherent material properties, but also by pipe design, construction and installation practices. For example, poor installation that results in the impingement of point loads on the pipeline will significantly magnify local stresses in the pipe wall. While these may be resisted by the current generation of PE grades, those earlier materials with poor slow crack growth resistance may fail. Alternatively, poor preparation of welded joints and fittings may result in premature failure. However, it is important to note that these types of failures reflect the quality of construction and installation practices rather than the performance of the PE material itself.

In some cases, water utilities have attempted to provide descriptions of failure modes that are observed (i.e., "circumferential fracture" and "longitudinal fracture"). While these descriptions are relatively clear in some cases, in others, they are not (i.e., "Blown section"). Although attempts have been made to explain these reported failure modes, failures are also often recorded simply as "pipe" or "leak." These latter descriptions do not provide any insight into the particular mode of failure that occurred.

FIELD PERFORMANCE OF PE PIPES IN AUSTRALIAN WATER UTILITIES

Table 2.1 lists the estimated annual failure rate (per length/per year) for those Australian water utilities that supplied data.

As shown, a wide range of failure rates was reported, from 5.9 for South Australia Water to 21.2 per 100km/per year for Ipswich Water. It is important to note that for all Australian water utilities who responded, the length of PE pipe in use is relatively small. Consequently, these reported average failure rates are merely estimates and should be interpreted carefully. It should also be noted that an average failure rate (per 100km/per year) was calculated by simply dividing the total number of recorded failures by the time period over which they were recorded and the *current* total length of PE pipe.

Table 2.1
Estimated annual failure rates for PE pipes in Australian water utilities

Utility	Current length of PE in network, km (miles)	Percentage of total network length	# failures	Recording period	Estimated annual failure rate per 100km/per year (per 100 miles/year)
City West Water	174.5 (278.4)	4.8	132	1996–2003	9.5 (15.8)
Yarra Valley Water	298.1 (478.0)	3.5	313	1996–2003	13.1 (21.8)
South East Water	301.7 (482.8)	3.9	258	1996–2003	10.7 (17.8)
South Australia Water	55.6 (89.0)	0.2	23	1995–2001	5.9 (9.8)
Ipswich Water	45.7 (73.1)	3.1	29	2003–2005	21.2 (35.3)

Table 2.2
Schematic diagram of pipe asset and failure data organization

Asset No.	Year laid	Age (years)									Length (m)
		0	1	2	…	10	11	…	…	40	
001	1980	x	x	x	x	x	1	2	0	2	101
002	1981	x	x	x	x	2	0	0	0	1	203
etc.	1983	x	x	…	…						

In practice, the total length of PE pipe and the number of failures will vary on a year-to year basis. Consequently, the annual failure rate will also change yearly. For a more accurate indication of annual failure rate (and how it changes with time), it is recommended that water authorities record and organize pipe asset and failure data in the format shown schematically in Table 2.2 (Jarrett et al. 2001).

By organizing pipe asset and failure data in this format, the changes in failure rates can be mapped accurately on a yearly basis. With sufficient failure data, this may also allow failure rates (/length/year) to be extrapolated into the future.

Breakdown of PE Failure Modes in Australian Water Utilities

For those authorities that supplied failure rate data, a breakdown of pipe failure mode was also requested. The distributions of failure modes are plotted below.

As shown in Figure 2.1, the majority of PE pipe failures (58%) reported by City West Water are given the description "Leak" and 29% are reported as "Burst." The remaining 13% of failures do not relate to PE pipes. While the majority of PE failures are attributed to leaks, it is unclear whether this description applies to pipes and/or joints. The breakdown of failure modes for Yarra Valley Water is shown in Figure 2.2.

As shown, the majority of failures (43%) are attributed to "Fitting failure." This general failure classification applies to any component other than the pipe itself. For example, failures of joints, valves, hydrants, ferrules etc. can all be given this description. However, further information

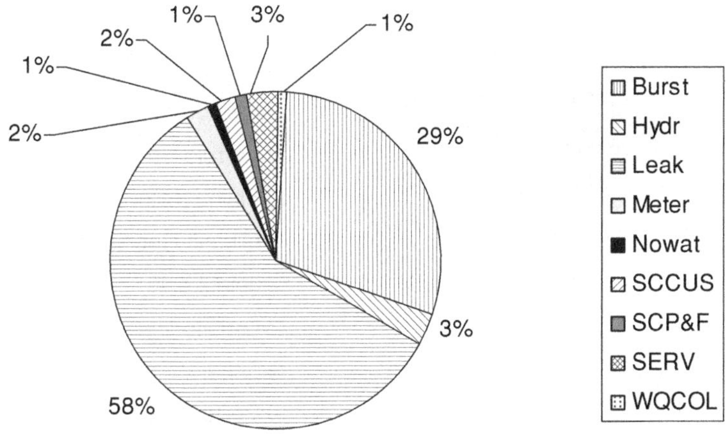

Figure 2.1 Breakdown of failure modes in PE pipes for City West Water (Australia)

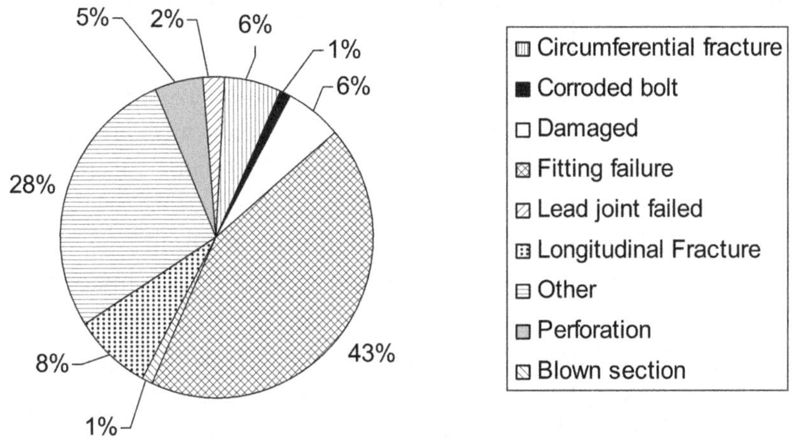

Figure 2.2 Breakdown of failure modes in PE pipes for Yarra Valley Water

is required to segregate these fitting failures. For example, the percentage of failures attributed to electro-fusion fittings and butt-welded joints are currently unknown.

Following fittings, 28% of PE failures are given the description "Other," and further information is required. 8% are described as "Longitudinal fracture." This failure mode may correspond to a single crack initiating and propagating axially along the length of the pipe and is more likely to have occurred in those early PE material grades, which exhibited relatively low slow crack growth resistance. The increased slow crack growth resistance of currently available PE material grades means that this failure mode is less likely to occur. Crack initiation in these cases is also often attributed to poor pipe installation practice resulting in point loads acting at the pipe outer surface.

6% of failures are attributed to "Circumferential Fracture," in which a crack initiates and propagates around the circumference of the pipe. Since the direction of crack propagation is perpendicular to the direction of maximum principal stress, this failure mode suggests that the pipe has experienced axial stresses, which can originate from a number of potential sources. For example, lateral ground movement around a poorly supported pipe can generate axial bending stresses. However, these stresses should be relatively low since the high flexibility of PE pipes

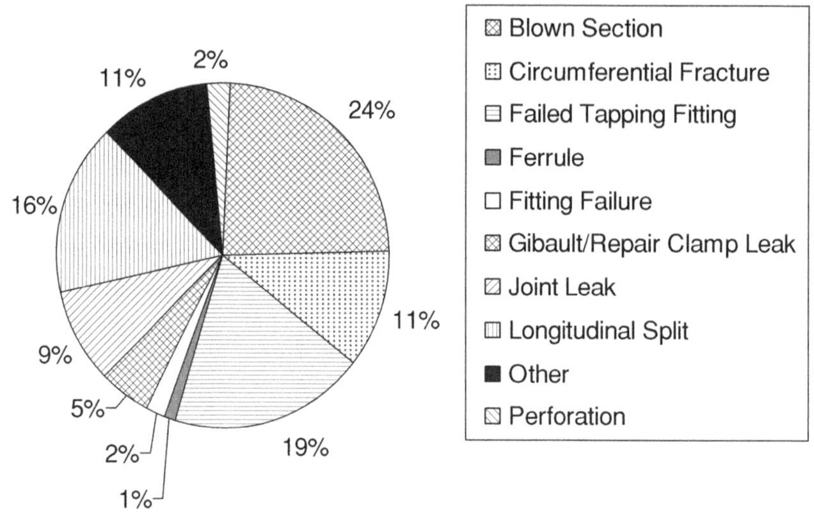

Figure 2.3 Breakdown of failure modes for South East Water

allows significant displacements to be tolerated without failure. It is possible that, in butt-welded PE pipes, these fractures can initiate at poorly formed welds and propagate circumferentially around the weld line. However, axial stress is still required to cause circumferential crack propagation. Other situations that may result in excessive local axial stress are sharp changes in geometry, such as the transition from a pipe to a socket fitting. It should be noted that again, these potential failure modes do not reflect the performance of the PE pipe material and can be avoided through proper jointing practice.

6% of failures are described as "Damaged," which corresponds to third party damage such as external impact from stray pick axes or backhoes during excavation. 5% of failures are described as "Perforations," which is a description normally reserved for through-thickness corrosion pits in metallic pipes. In practice, perforations in PE pipes may be very small punctures, created by slow crack growth through the pipe wall.

The breakdown of PE failures for South East Water is shown in Figure 2.3.

For South East Water, 24% of PE failures are given the description "Blown Section." This corresponds to a pipe failure when a section of the pipe wall has been removed. This type of failure could possibly occur when a brittle fracture bi-furcates (i.e. a second crack initiates) and then re-joins the original crack path. The resulting bounded section of the pipe wall is then free as shown schematically in Figure 2.4.

However, this brittle failure mode is thought to be extremely unlikely in the current generation of PE material grades. While this mode may be possible in earlier PE grades, it is still unlikely. A more probable explanation for this type of failure mode could be external impact (a stray pick axe or backhoe for example) from adjacent civil works. As such, it may be better classified as "damaged" as used by Yarra Valley Water in Figure 2.2.

After blown section failures, 19% of PE failures are described as "Failed tapping fitting." For metallic fittings, this failure could occur by corrosion in aggressive soil conditions. Alternatively, poor surface preparation leading to inadequate fusion at the pipe-fitting interface can result in failure in electro-fusion tapping fittings. As stated above, these failures reflect inadequate joint preparation practices rather than failure of the PE pipe system itself. Following tapping fittings, 16% of PE failures recorded by South East Water are reported to occur by longitudinal fracture, as described above.

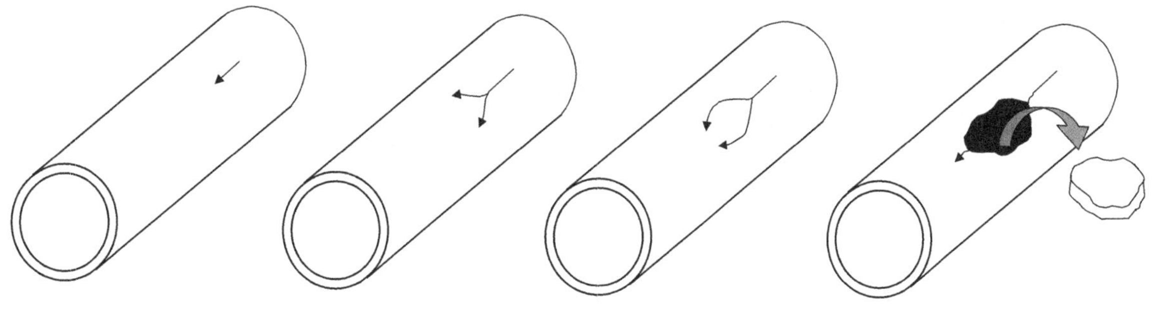

Figure 2.4 Schematic illustration of how a blown section can occur in PE pipe

Table 2.3
Average failure rates (per 100km) from UKWIR national failure database (adapted from MacKellar and Pearson, 2003)

Average failure rate per 100km	Year					
Material group	1998	1999	2000	2001	2002	Average
Asbestos cement	16.4	17.1	15.1	15.8	15.6	16
Ductile Iron	5	5.3	4.8	4.8	6.5	5.28
Iron	23.7	23.7	19.1	21.7	12.3	20.1
PE	**3.5**	**2.9**	**3.3**	**3.1**	**3.0**	**3.16**
PVC	9.6	9.1	7.2	7.4	3.3	7.32
Steel	5	6.1	5.8	5.7	33.1	11.14
Unknown	0.1	0	0	0.1	15.9	3.22

11% of failures are described as "Circumferential fracture" and "Other" respectively. 9% of failures are attributed to "Joint leak," which may be caused by leakage from bolted flange joints or butt-fusion welds. Again, the type of joint used in the field should be determined.

FIELD PERFORMANCE OF PE PIPES IN UK WATER UTILITIES

For comparison with the current field performance survey, previously recorded failure data for PE pipes in UK water utilities was made available by UK Water Industry Research (UKWIR). As part of a UK Water Industry Research (UKWIR) research project, MacKellar and Pearson (2003) compiled pipe asset and failure data from a total of 17 UK water utilities. Results from this study were used to compile a nationally-agreed failure database for water pipes in UK utilities (MacKellar and Pearson, 2003). An overview of the UKWIR national failure database is presented in Table 2.3 (MacKellar and Pearson, 2003).

As shown in Table 2.3, a range of average failure rates has been reported for different pipe materials. The highest in this range is 20.1 per 100km/per year for Cast Iron, and the lowest in the range 3.16 per 100km/per year for PE pipes.

Table 2.4
Failure modes in PE pipes
(adapted from MacKellar and Pearson, 2003)

Failure mode	Percentage of total PE pipe failures (%)
Pipe	30
Joint	19
Circumferential	4
Longitudinal	3
Pin Hole	2
Ferrule	1
Electrofusion	0
Other	3
Unknown	38

Breakdown of PE Failure Modes in UK Water Utilities

The range of different failure modes in PE pipe is given in Table 2.4.

As shown, the majority of failures were reported as "unknown," which highlights the need for consistent failure reporting in the UK. This has been addressed as part of the UKWIR study (MacKellar and Pearson, 2003) and a protocol for collection and reporting of failure data has been proposed. 30% of reported failures were given the description "pipe," which suggests that a pipe failure occurred as opposed to a joint or fitting failure. However, it is unclear what type of pipe failure occurred and again, the need for useful failure descriptions is clear. Anecdotal evidence from the UK (Beech, 2006) suggests that the majority of these failures are caused by third party damage (such as that from adjacent excavation works) and do not reflect PE pipe failure under normal service loads. It has also been suggested that many of these "pipe" failures are actually joint or fitting failures which have been incorrectly described (Beech, 2006). 3% of failures are reported as "circumferential" (circumferential fracture) and 3% are reported as "longitudinal" (longitudinal fracture). 2% of failures are reported as "pin hole," which may correspond to very short longitudinal fractures. 19% of recorded failures have been attributed to joints, which (as with the Australian experience) are thought to reflect inappropriate preparation and installation of joints rather than failure of a PE pipe under normal service conditions (Beech, 2006).

Average Failure Rate vs. Age Curves for PE Pipe in UK Water Utilities

In addition to the written report summarizing recorded PE pipe failures in the UK (MacKellar and Pearson, 2003), UKWIR also supplied the UK national database of mains failures in Microsoft Access form. An example of the type of data included is shown in Figures 2.5 and 2.6.

As shown, the UKWIR database is organized into a "Mains Summary" and a "Burst Summary." The Mains summary contains information describing the diameter, material installation year for different materials. The type of surface and length of main installed is also specified. The Burst Summary contains information describing individual failures in pipes. As shown in

Microsoft Access - [UKWIR Mains Summary : Table]

Diameter Group	Material Group	Material Sub Group	Period Laid	Surface Type	Surface Sub Type	Sum of Length (m)
200	PE	PE100	2000			1049
200	PE	PE80	1975	Unknown	Unknown	6
200	PE	PE80	1980	Footpath	Footpath	4
200	PE	PE80	1980	Highway	Highway (Light Use)	3
200	PE	PE80	1980	Unknown	Unknown	4282
200	PE	PE80	1985	Footpath	Footpath	3
200	PE	PE80	1985	Unknown	Unknown	2988
200	PE	PE80	1990	Field	Field	740
200	PE	PE80	1990	Footpath	Footpath	510
200	PE	PE80	0	Unknown	Unknown	26258
100	PE	PE	1865	Unknown	Unknown	50
400	PE	PE	1985	Unknown	Unknown	6123
400	PE	PE	1980	Unknown	Unknown	1375
400	PE	PE	0	Unknown	Unknown	6538
100	PE	PE	0	Field	Field	4
100	PE	PE	0	Highway	Highway (Heavy Use)	11
100	PE	PE	0	Highway	Highway (Light Use)	13

Figure 2.5 Extract from UKWIR national database: Mains summary

Microsoft Access - [UKWIR Bursts : Table]

Burst Type	Burst Sub Type	Burst Month	Burst Year	Diameter Group	Material Group	Material Sub Gr	Period Laid
Pipe	Pipe	5	2001	50	PE	PE	1955
Pipe	Pipe	7	2001	50	PE	PE	0
Joint	Joint	9	1999	50	PE	PE80	1940
Pipe	Circumferential	2	2000	50	PE	PE80	1985
Pipe	Pipe	5	2001	50	PE	PE	0
Pipe	Pipe	2	2001	50	PE	PE	0
Other	Other	9	1999	50	PE	PE80	1940
Joint	Joint	8	2000	50	PE	PE80	1950
Pipe	Pipe	1	2001	50	PE	PE	0
Pipe	Pipe	2	2001	50	PE	PE	1985
Pipe	Circumferential	11	1999	50	PE	PE80	1995
Pipe	Pipe	2	2001	50	PE	PE	0
Unknown	Unknown	9	1998	50	PE	PE80	2000
Joint	Joint	4	1999	50	PE	PE80	1920
Pipe	Pipe	4	2001	50	PE	PE	0
Unknown	Unknown	1	1998	50	PE	PE80	1970
Unknown	Unknown	5	1998	50	PE	PE80	1995
Pipe	Pipe	2	2001	50	PE	PE	0
Pipe	Pipe	7	1997	50	PE	PE	0

Figure 2.6 Extract from UKWIR national database: Burst summary

Figure 2.6, the type of failure is given along with the month and year that the failure was recorded. The pipe diameter, material and installation year is also supplied.

Combining the information on the Mains and Burst summaries, the effect of age on average failure rates for UK PE pipes can be determined. An average failure rate (per pipe length/year) can be defined for a particular pipe age as (Jarrett et al. 2001):

$$\text{Average failure rate at Age A} = \frac{\text{Number of observed failures at age A}}{\text{Exposed pipe length at age A}} \tag{2.1}$$

For a particular pipe age, the exposed pipe length is that length of pipe that is the same age at the time of failure (Jarrett et al. 2001). An algorithm to calculate average failure rates for pipes of different ages in the UKWIR database has been written using the Python 2.3 programming

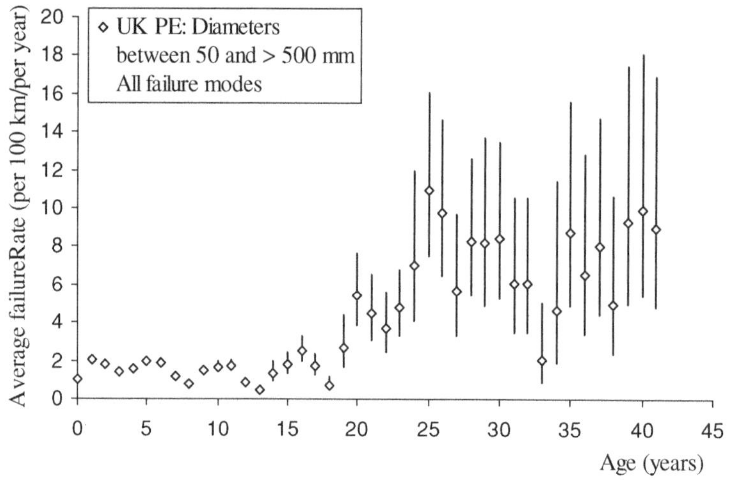

Figure 2.7 Average failure rate vs. age for UK PE pipes: All failure modes in diameters between 50mm and > 500mm

language and is listed in appendix A. The resulting graph of average failure rate (per 100km/per year) vs. age for PE pipes in the UKWIR database is given in Figure 2.7.

As shown, average failure rate increases at a very low rate with pipe age (from 2 per 100/km/per year, to 4 per 100km/per year) until an increase occurs for pipes greater than 20 years old. The 95% confidence limits associated with each estimate of average failure rate is approximately given by (Jarrett et al. 2001)

$$\text{Average failure rate} \; \overset{\times}{\underset{\div}{-}} \left[\exp\left(\frac{2}{\sqrt{n}} \right) \right] \tag{2.2}$$

where n is the number of observed failures that comprise the estimated average failure rates. Since the installed length (and hence exposed length) of these older pipes is relatively low, fewer failure events have occurred and the error associated with these 95% confidence limits average increases. However, the observed increase in average failure rate falls within the limits set by this error, suggesting that the effect of pipe age is valid.

To assess the influence of failure mode on this observed influence of pipe age, Figure 2.8 shows average failure rates obtained from those events given fracture descriptions only (i.e., "longitudinal," "circumferential," and "pin hole" fractures).

As shown in Figure 2.8, the trend in fracture failures with pipe age is similar to the trend observed for all failure modes. Again, a relatively constant average failure rate is observed until an increase occurs for pipes aged around 20 years old. A decrease in failure rate is also observed for pipes aged 34 years old. One possible reason for this decease is an improvement in pipe installation practices for this particular set of pipes. This may have reduced the number of point loads acting on buried PE pipelines.

Referring back to chapter 1, the development of pipe grade PE materials in Europe produced second generation MDPE materials in the 1970s and third generation, bimodal HDPE materials from 1990 onwards. Consequently, it is likely that a combination of PE material grades is included in the failure rate vs. age graph in Figure 2.8. In an attempt to isolate the effect of

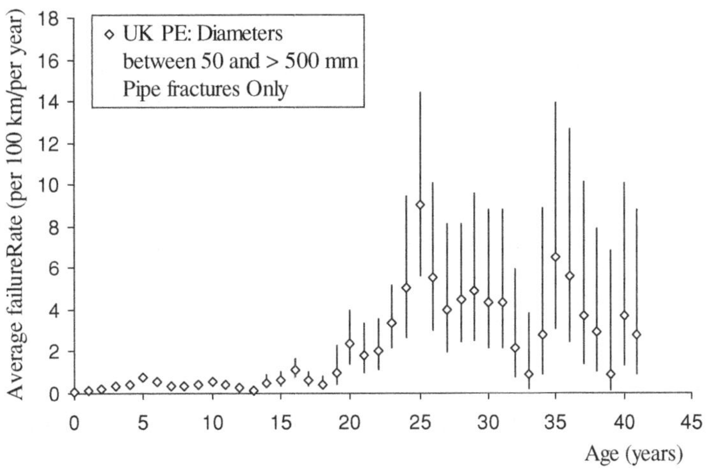

Figure 2.8 Average failure rate vs. age for UK PE pipes: Fracture failure modes only in diameters between 50mm and > 500mm

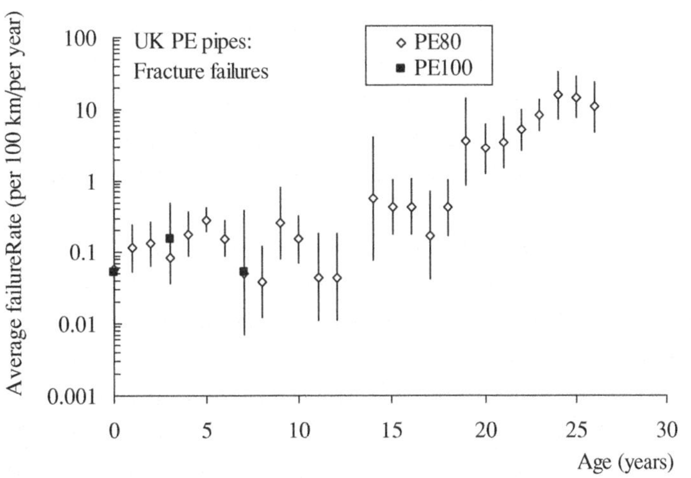

Figure 2.9 Average failure rate vs. age for UK PE pipes: Fracture failure modes only in pipes installed between 1970 and 1990 and post-1990

different PE grades on failure rates, a comparison was made between those pipes with material sub types listed as "PE80" in the UKWIR database and those with material sub types listed as "PE100."

Figure 2.9 compares average failure rate vs. age data for these two pipe categories. Again, only those failures given a fracture descriptor have been extracted from the UKWIR database for each material type. Since failure rates in younger pipes are very low, a logarithmic scale is used.

As shown in Figure 2.9, failure rates for "PE100" pipes are relatively low at approximately 0.1 per 100km/year for pipes aged up to 7 years old. Due their recent installation years, no failure rate data has been recorded in the UKWIR database for older "PE100" pipes. In comparison, average failure rates in "PE80" are also around 0.1 per 100km/year for pipes aged up to

12 years old. The average fracture failure rate then increases up to 18 per 100km/year for 25-year old pipes. While this may indicate a genuine effect of age on the fracture failure rate in "PE80" pipes, it may also reflect poor installation practices for older pipes that have since been improved. Furthermore, although failure descriptions are restricted to pipe fracture (i.e., "Longitudinal," "Circumferential," and "Pin Hole"), anecdotal evidence suggests that a large of these fracture failures may actually have been caused by third party damage (Beech, 2006). Consequently, the increasing average failure rate in Figure 2.9 may not be a genuine effect of age.

It should also be noted that (as stated in the introduction to this chapter) no further information concerning the particular generation of PE material was available in the UKWIR database. Consequently, it is likely that a range of different PE generations with different slow crack growth resistances are present in the dataset.

FIELD PERFORMANCE OF PE PIPES IN US/CANADIAN WATER UTILITIES

A total of 55 US/Canadian water utilities were surveyed for field performance information on PE pipes in their network. The list of water authorities thought to use PE pipes was supplied by AwwaRF, and (where available) e-mail contact addresses were obtained from web pages for each utility. From the 55 utilities surveyed, a total of 33 responded to a request for failure data in PE pipes. However, very little quantitative data for length, age and failure history of PE pipes was supplied, and the majority of responses gave anecdotal evidence describing the experience of each utility with PE pipes. The results of the survey are given in Table 2.5.

As shown in Table 2.5, some entries suggest that previous problems occurred with early PE materials, which, as discussed previously, exhibited poor resistance to slow crack growth. However, anecdotal evidence from industry indicates that these failures were also influenced by:

- Point load impingement in "stoney" soils
- Excessive bending from poor installation and differential soil settlement
- Sharp changes in geometry at improperly dimensioned insert fittings
- One particular batch of PE pipe produced from a single supplier in 1974 which was gradually embrittled under UV due to inadequate stabilizer levels.

Based on these, and other findings relating to properties that affect quality of field performance, a number of important changes were made to AWWA and ASTM standards in the mid-1980s. These included a requirement for stabilizer protection quality; a minimum requirement for stress-crack resistance and a "bend-back" test for the confirming of high ductility at the inner surface of a PE pipe. Together with other changes, these resulted in improvements in the overall performance of PE piping. Therefore, (as with the Australian and UK performance data), these failures in early PE grades reported by US utilities are not representative of the current situation of PE materials with dramatically increased resistances to slow crack growth and significantly improved pipe installation practices. This observation is borne out by those entries in Table 2.5 which report zero failures in recently installed larger diameter HDPE water supply pipe.

In addition to the improvements reported above for PE water pipes, the US gas distribution industry has conducted various surveys regarding the field experiences that have been reported with PE piping. These surveys have greatly contributed to improvements of pipe materials, material and pipe specifications and construction practices. The result has been a steadily improved field performance with PE becoming the dominant material for gas distribution pipelines.

Table 2.5
Responses from field performance survey for PE pipes in US/Canadian water utilities

Utility name	PE services	PE water supply	Comments
Aqua Pennsylvania, Inc.	No info	Yes	Small percentage of network length is HDPE pipe. No failure problems reported.
Boston Water & Sewer	No	No	PE not currently used
City of Anaheim	Yes	No	PE not currently used for larger diameter water supply.
City of Asheville	No	No	PE not currently used
City of Austin	Yes	No	Limited use of PE for small diameter service connections.
City of Cincinnati Water Works	No info	Yes	One project using HDPE pipe, installed in spring 2003. Approximately 2,500 feet, 8 inch SDR 11, and no failures reported to date.
City of Cleveland	No info	Yes	PE used in a small number of speciality locations.
City of Gallup	No	No	PE not currently used, but will probably be used in the future.
City of Norfolk	Yes	Yes	Three relatively short sections (between 350 feet and 625 feet) of HDPE pipe used for larger diameter water supply. Directionally drilled creek crossings in each case.
City of Portland Water Bureau	No	No	Slip lining trials with HDPE have been conducted using pipes in the yard, but no projects in the field to date.
City Water, Light & Power	No	No	PE not currently used.
Daly City	Yes	No	0.75" and 1.5" PE service connections previously installed. Connectors prone to brittle failures. Rocks puncturing PE services.
Denver Water	No	No	PE not currently used.
Des Moines Water Works	No info	Yes	Approximately 1 mile of PE in system and no failures reported to date.
EPCOR Water	No info	Yes	5.8km of PE used in larger diameter water distribution network. No failures reported to date.
Greenville Utilities Commission	Yes	No	Used PE for 0.75" and 1" service connections for 25 years then stopped 5 years ago due to number of reported leaks.

(continued)

Table 2.5 (Continued)

Utility name	PE services	PE water supply	Comments
Henrico County	No	No	PE pipe not currently used.
Louisville Water Company	No info	Yes	Between 6 and 7 miles of HDPE water supply pipe installed with no failures to date.
Marshalltown water	No	No	PE pipe not currently used.
Orlando Utility Commission	Yes	Yes	Previous premature failures in lower density PE service connections have "soured" experience with PE. HDPE used on a limited basis for larger diameter water supply and only when there is a construction advantage. Thought that PE pipes do not locate well with existing technologies and are susceptible to construction damage.
Ottawa Water	Yes	Yes	No failures in large diameter PE water supply pipes have occurred since 1997.
San Jose Water Co.	Yes	No	PE used extensively for 0.75" and 1" service connections. No failure information provided to date.
Seattle Water Dept.	Yes	No	Early brittle failures in PE service connections from external loads such as soil movement and adjacent excavation.
St. Louis County Water Co.	No info	Yes	Larger diameter PE water supply pipe used on a limited basis while exploring directional drilling techniques.
Tacoma Public Utilities	No info	No	PE not currently used.
United Water Toms River	Yes	No	Changed to PE service connections from copper services between 1974 and 2001. Failures reported in PE service connections.
United Water Delaware	Yes	No	Similar experience to Toms River, with PE also being damaged by stone penetration. PE still used for service connections since replacement costs are relatively low.
Windsor Utilities Commission	No info	No	PE not currently used.

According to industry expert opinion, lessons learned from the development of PE for gas applications have also been applied to the development of standards and practices that apply to PE water pipes.

In a recent study of the field performance of PE gas pipes Palermo (2004) presents average failure rates for a particular batch of Aldyl "A" PE2306 pipes and fittings which were thought to be susceptible to brittle cracking. For these LDIW (Low Ductile Inner Wall) Aldyl "A" PE2306 pipes and fittings (manufactured between 1971 and 1972), an average failure rate of 0.04 leaks per mile/per year is quoted. For "improved Aldyl A" (manufactured between 1974 and 1984) the average failure rate is 0.016 leaks per mile/per year. However, these failure rates also include those events occurring by rock impingement which reflects poor installation practices rather than actual pipe/fitting failure rates. Removing rock impingement failures from the data (and expressing rates as events per 100km/year) results in failure rates of 1.98 failures/100km/year for Aldyl "A" and 0.9 failures/100km/year for "improved Aldyl A." For comparison, average failure rates for Australian PE *water* pipes are 8.9 failures/100km/year and 3.16 failures/100km/year for UK PE *water* pipes. Clearly, the average failure rates in US PE gas pipes are lower than this, which would indicate superior performance. However, it is also important to note that these quoted failure rates for US PE are for *gas* pipes which would operate under lower internal pressures than *water* pipes.

STANDARD FAILURE CLASSIFICATIONS

The survey carried out in this study shows that there is considerable variation in the way that water utilities report their failure data for PE pipes. A standardized methodology would make comparison of failure information between water utilities much easier. Recently, the Water Services Association of Australia (WSAA) has recommended that a standardized reporting methodology be used by its members for all pipes types, and the adoption of a similar process by water utilities in the USA would make analysis of failure data to predict long-term failure trends much easier. This methodology is defined in Table 2.6.

For joint and fitting failures, recording the type of joint/fitting is also important. For example, distinctions should be made between welded joints (butt fusion and electrofusion) and mechanical joints. Furthermore, those failures that occur due to poor surface preparation and subsequent inadequate fusion should also be isolated. Including these events in average failure rate figures for PE pipes may lead to a falsely high failure rate and unfairly represents that field performance of PE pipes. For those pipe fractures that do occur, additional information to distinguish between failures under normal service loads and those that occurred under third party damage is also needed. The "damaged" description used by Yarra Valley Water in Figure 2.2 is recommended as a secondary description to follow the type of fracture.

Further information is also required to identify different PE material grades and hence isolate those pipe fracture failures that occurred in early materials with lower slow crack growth resistance. Although PE pipes are given the sub description "PE80" and "PE100" in the UKWIR database, it is possible that even within this description, a wide range of different slow crack growth resistances will exist.

CONCLUDING REMARKS

This chapter has presented responses to a water industry survey of the field performance of PE water pipes. While average failure rates and failure modes were supplied by some water

Table 2.6
WSAA terminology for reporting failures in water distribution pipes

CSIRO/WSAA designation	Failure mode to be grouped under this description
Blown section	Burst, burst main, burst section, broken section, blown section
Circumferential fracture	Broken back, circumferential fracture, circumferential split
Ferrule	Ferrule replacement, saddle and ferrule, other (specify)—ferrule, ferrule out of main, split in ferrule, leaking ferrule, ferrule out, leak on ferrule, supply court, fracture at ferrule, split ferrule
Fitting failure	Leaking washout, faulty fittings, fitting split, leaking fitting, leak on tee, fitting failure, hole in bend, corroded fitting, washout, blank end blow off, replace defective T, broken fitting, faulty fitting, blow out on bend, blown P-bend
Gibault/repair clamp leak	Bolts on gibault corroded, gibault/repair clamp leak, leak on clamp, blown gibault rubber, bolts on gibault corroded, leaking gibault joint, broken fitting gibault coupling, corroded bolts, faulty SS clamp
Joint leak	Blown rubber on flange, blown joint, lead joint, joint leak, leak on collar, leaking rubber ring
Longitudinal split	Split on bend, split in main, split pipe, longitudinal split, longitudinal fracture, spiral fracture, split, split main
Other	Blank ends, no information, job not finished, blown T, other, damaged, not stated, poor supply, fire plug riser split, ball valve, end blown off
Perforation	Perforation, pinhole in pipe (note that PVC and PE perforations may be splits)
Failed tapping fitting	Tapping band, leaking saddle, saddle, tapping section, tapping saddle, broken band, blown saddle, blown tapping saddle, broken saddle
Unknown	No type entered

utilities, it is essential to note that in addition to material properties, field performance is influenced by pipe construction and joint installation practices. For example, failures that are attributed to joints and fittings in PE pipelines are likely to be caused by a number of factors relating to preparation and installation such as improper surface preparation in joints and fittings (i.e., misalignment, dirt contamination, absence of a stiffening insert for compression fittings). Such factors do not reflect the true performance of a pipe material. Anecdotal evidence also suggests that in many cases, reported pipe fracture failures may actually be attributed to poor pipe installation resulting in point load impingement at the pipe outer surface. Such failures are preventable and should also be separated from others that occur under normal service loads since they give a false representation of PE pipeline performance.

The current failure descriptions used by water utilities are also inadequate and do not provide any insight into the cause of failure. More detailed descriptions are required to correctly identify those failures which occur due to poor installation and jointing practices. They can then be treated separately as representing the current state of pipe installation practice rather than the current performance of PE pipe materials. A further complication in interpreting field performance data is the different PE material grades that will be present in any dataset. Early generations of PE material may exhibit low slow crack growth resistance and as such, may be susceptible to pipe fracture. However, due to the significantly improved slow crack growth resistance, pipes manufactured from currently available PE grades means would not be expected to fail by fracture under normal operating loads. Regardless of these problems interpreting the cause of PE pipe failures, the data presented in this chapter indicate an overall failure rate that is relatively low compared to other materials used for water pipelines.

Field Performance of PE Water Pipes in Australia and the UK

Of the Australian respondents to the field performance survey, Ipswich Water reported the highest average failure rate at 21.2 per 100km/per year and the lowest failure rate of 5.9 per 100km/per year was reported by South Australia Water. However, it should also be noted that the length of PE pipe in all Australian water utilities was relatively small. Consequently average failure rates are merely estimates and should be interpreted very carefully. A more rigorous indication of the field performance comes from PE pipes in the UK, and is based on data from the UKWIR national database (MacKellar and Pearson, 2003). Analysis of the UKWIR data indicates an average failure rate of 3.16 per 100km/year for PE pipes, which is the lowest of all materials in use (MacKellar and Pearson, 2003).

Examination of the UKWIR database indicates that 4% of PE pipe failures are described as "Circumferential," 3% as "Longitudinal," and 2% as "Pin Hole." While each of these descriptions is intended to describe pipe fracture failure, industry expert opinion suggests that, due to their high flexibility, circumferential fracture failures are not expected to occur in PE pipes (Mruk, 2006). A more plausible explanation for this failure mode is crack initiation from a poorly formed butt fusion weld. Anecdotal evidence from the UK suggests that the majority of failures that are recorded in pipes can actually be attributed to third party damage rather than failure of the pipe under normal service loads (Beech, 2006). 19% of failures in the UK are described as "Joints" but no further information describing the type of joints that fail. Again, evidence from the UK (Beech, 2006) suggests that these failures can be attributed to poor preparation and installation practice rather than joint failure under service loads.

Field Performance of PE Water Pipes in the US

From a total of 55 US/Canadian water utilities surveyed for field performance information on PE pipes in their network, 33 responded to a request for failure data in PE pipes. The majority of responses gave anecdotal evidence describing the experience of each utility with PE pipes. Those water utilities whose networks contained PE service connections reported that:

- PE pipes have been used previously in US/Canadian water utilities for small diameter (0.75 inch to 1.5 inch) service connections.
- In some cases, use of these PE service connections was discontinued due to large numbers of premature failures.
- These premature failures occurred by brittle fracture and were attributed to point loading from surrounding backfill, damage from adjacent construction or a reported loss of pipe flexibility.

It is important to distinguish previously reported failures in these small diameter service connections from the expected field performance in larger diameter water supply pipes manufactured from currently available PE materials in the US. Not only have pipe installation practices improved over time, but (as discussed in chapter 1) currently available PE materials offer significantly increased slow crack growth and fracture resistance properties compared to these previous materials used for small diameter service connections. This is reflected in the response of those water utilities who had recently installed large diameter PE pipes for water supply:

- HDPE pipes for larger diameter water supply have been installed in relatively short lengths (i.e., less than 10 miles) and is relatively young (i.e., less than 12 years old).
- No failures have been reported in these larger diameter HDPE water supply pipes to date.
- High Density Polyethylene (HDPE) has been used for larger diameter water supply (i.e., 8 inch and 12 inch diameter pipes) when there is a significant construction advantage, such as that provided by directional drilling.

In agreement with this anecdotal evidence, Sandstrum (2004) and Rubeiz (2004) document a series of US/Canadian case studies where the flexibility of HDPE pipes was reported to offer construction advantages in difficult installation environments.

CHAPTER 3
FAILURE MECHANISMS IN POLYETHYLENE PIPELINES

INTRODUCTION

The field performance survey presented in chapter 2 indicated that the average failure rates in PE water pipes are low compared to other pipe materials used for water transportation. Descriptions of those failures that do occur in service were also unclear and are thought to reflect inadequate jointing and pipe installation practices rather than the performance of the pipe material itself. A more detailed description of joint and fitting failures which are caused by poor preparation is given in this chapter. While the majority of PE failures are caused by inadequate jointing, pipe fracture failures can occur in older material that that exhibit relatively low crack growth resistance. This chapter describes the mechanisms of field failures in more detail. It should be noted that in comparison, new generation PE materials are highly resistant to cracking under normal operating conditions.

JOINT/FITTING FAILURES IN PE SYSTEMS

As discussed in chapter 1, PE pipelines contain either butt-welded joints between pipes and/or electrofusion joints between pipes and fittings, such as tees and couplings. While these joints should retain the structural integrity of the pipeline if they are formed correctly, poor preparation of the surfaces to be joined can result in premature failure in service. For example, Figure 3.1 shows one possible failure mode that can occur in poorly installed PE electrofusion joints and fittings. The sample was provided by the UK Water Research Centre (WRc) and the view in Figure 3.1 shows a longitudinal section through a PE pipe and coupling. The pipe and coupling thicknesses are highlighted.

This failure was originally observed as a large "bubble" at the inside surface of the pipe. When the sample was cut open as shown in Figure 3.1, the bubble was revealed as a large separation between the inside surface of the coupling and the outer surface of the pipe. The distortion of the pipe and coupling suggests that considerable pressure has been created within the joint to cause separation. This failure mode is most likely the result of entrapped moisture between the pipe and coupling prior to fusion. Under high fusion temperatures, this moisture would have boiled to steam and expanded, creating the bubble in the molten polymer. The steam would have forced its way into the thin layer of polymer between the electrical elements in the coupling and its inner surface. Separation has occurred within the coupling, and behind the electrical element.

The sample in Figure 3.2 shows a slightly different type of failure mode that has been previously observed in electrofusion couplings used by UK water utilities. This sample was also provided by WRc UK.

In this case, molten polymer has flowed from the interface between the pipe outer surface and the coupling inner surface. While this failure may also have been caused by entrapped moisture between the pipe and the coupling, it is also possible that the molten polymer flow was due to excess ovality of the pipes that were to be joined. For relatively small diameters, PE pipes are supplied in coils which may exhibit ovality in cross section. If this ovality is extreme, large gaps between the coupling inner surface and pipe outer surface can allow molten polymer to flow out of the joint (Figure 3.2). Guidelines in the UK suggest that if ovality is observable, the sections of

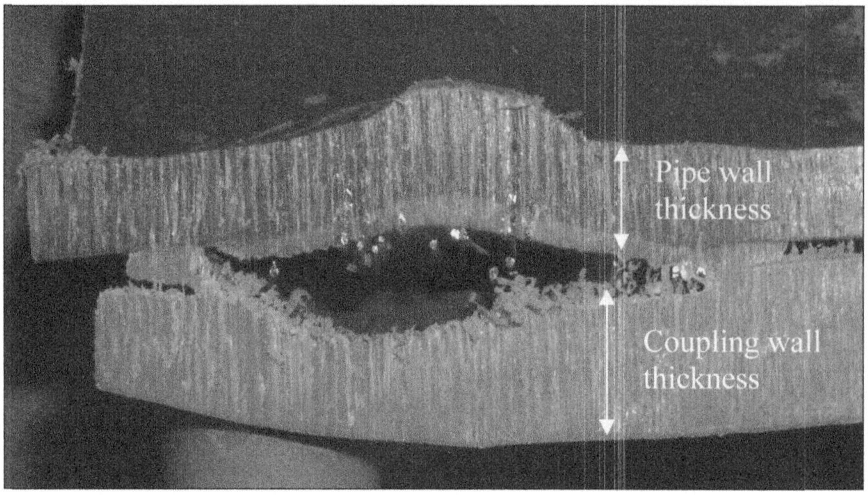

Figure 3.1 Separation of electrofusion joint in PE pipe—through-thickness view

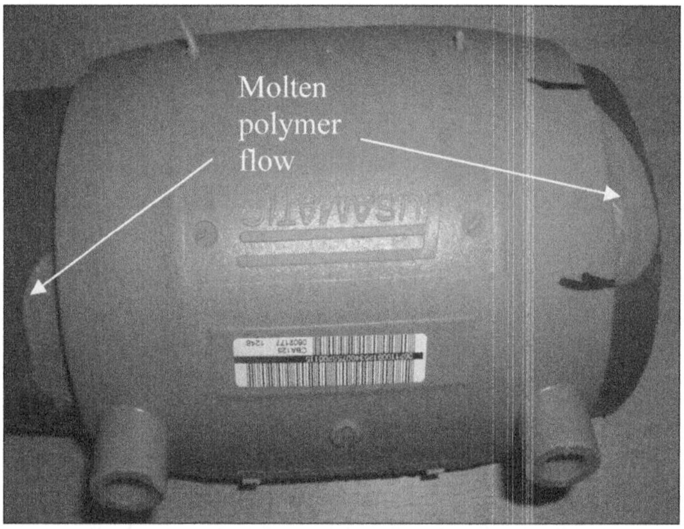

Figure 3.2 Molten polymer failure of electrofusion coupling

pipe to be joined are re-rounded prior to electrofusion jointing. It should be noted the failures in Figures 3.1 and 3.2 are attributed to poor practice. Standard practices, such as ASTM F1290 ("Standard Practice for Electrofusion Joining Polyolefin Pipe and Fittings"), and guides offered by electrofusion fitting manufacturers give specific guidance for avoiding problems such as the ones that are featured here.

Another failure mode that can occur from poor preparation of electrofusion couplings is separation due to contamination and pitting of the pipe/coupling interface. The photograph in Figure 3.3 shows pitting and contamination at the inside surface of a failed electrofusion coupling, supplied by WRc UK.

The joint between the pipe and the coupling could be forced apart very easily with a hammer and chisel. This kind of brittle failure indicates very poor adhesion at the pipe/coupling interface. As shown in Figure 3.3, the jointing surfaces of both the coupling and the pipe show

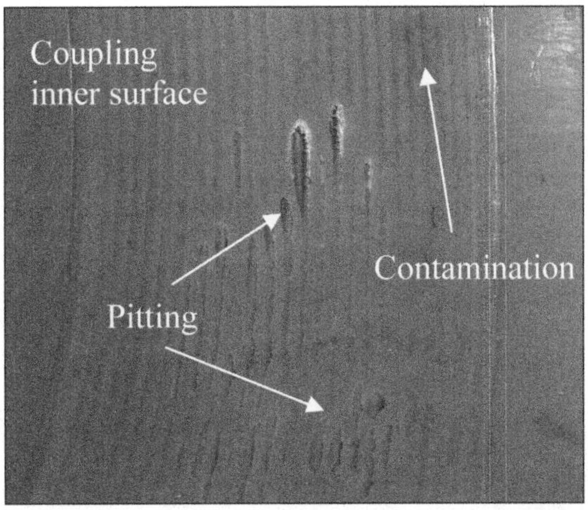

Figure 3.3 Contamination and pitting of electrofusion coupling—inside surface view

areas of pitting and the heating wires are visible in places. Similar to the previous failures, this indicates of the presence of moisture and subsequent steam bubbles in the joint at the time of fusion. Areas of dirt contamination are also visible, indicating poor cleaning and preparation of the pipe outer surface prior to electrofusion.

RAPID CRACK PROPAGATION IN PE PRESSURE PIPES

Rapid Crack Propagation (RCP) was identified as a potentially catastrophic failure mode in high pressure steel gas pipelines in the 1970s. Impact testing of these pipes revealed that although crack initiation was difficult, once it was achieved, uncontrolled crack propagation could be sustained at relatively low operating pressures. The arrival of second-generation PE materials in the UK in the early 1970s prompted British Gas to implement a design code for their larger diameter pipes, acknowledging the possibility of this type of failure occurring at lower operating pressures. RCP describes a crack traveling along a PE pipeline for extensive distances, at speeds in excess of 900 ft/second. For gas pipelines, the rapid escape of compressed gas in such an event can cause severe structural damage, and in the case of a fuel gas line, the potential for disaster is clear. Apparently since the 1980s several RCP events have occurred, although none are well documented. According to industry expert opinion, these previous RCP events occurred in materials which are now recognized as having poor RCP resistance, operating at an internal pressure considerably higher than the intended pressure rating. However, it is generally recognized that crack initiation cannot be prevented completely (from a badly aimed pick for example) and a design philosophy based on the prevention of RCP should be adopted.

Although RCP is theoretically possible in water-pressurized PE pipes it is even less likely to occur in service than in gas pipes. As described by Greenshields (1997), the rapid decompression of pressurized water upon crack initiation removes the available driving force for crack propagation. Consequently, only very short crack lengths (less than 1 diameter) are obtained during RCP of 100% water-pressurized PE pipe. However, the inclusion of occluded air pockets in water-pressurized pipes may increase crack propagation lengths (Greig, 1988). It should be noted that no known incidents of RCP have occurred in PE water pipes in service.

Figure 3.4 Slow crack growth failure in a 50mm diameter PE pipe. Installed in South Australia in 1994, failed 2004.

SLOW CRACK GROWTH IN PE PRESSURE PIPES

An inspection of asset failure databases in chapter 2 indicates that the proportion of joint/fitting and pipe fracture failures that occur varies between water utilities. A major contribution to this variation is the nature of the PE material in use. Specifically, variations in Slow Crack Growth (SCG) resistance are important and vulnerability SCG has been determined to be the major determinant of the potential life of a buried PE piping system under actual service conditions. As an illustrative example of a SCG failure in service, the photograph in Figure 3.4 shows a failure that was reported in a 50mm (2 inch) diameter PE pipe, installed in South Australia in 1994.

Leakage from the resulting fracture can be seen towards the right hand side of Figure 3.4. The fracture is relatively short (approximately 7mm (0.3 inches) long at the pipe outer surface) and may therefore have been reported as a "pin hole" failure. The presence of the repair clamp to the left of the fractured area suggested that similar short fractures had occurred previously, indicating that the material exhibited a relatively low resistance to SCG. This was verified when the sample was exhumed and the clamp removed, to reveal an additional fracture of similar length. Relatively large rocks can also be observed in the soil surrounding the pipe. However, it is unclear whether these rocks exerted point loads on the pipe prior to failure, or were simply exposed during excavation.

The photograph in Figure 3.5 shows the fracture surface created by this failure. The through-thickness view is shown and the pipe wall thickness is approximately 10mm.

The first points of interest are the elliptical striations on the surface which emanate from a point on the pipe inner surface. This is evidence of crack initiation from a defect at the pipe inner surface followed by radial growth (with striation fronts expanding) through the pipe wall to the outer surface. The striations indicate intermittent crack growth and are left behind when a crack stops temporarily during growth. Similar striations are observed during fatigue crack growth under cyclic loads. However, this "stop-start" crack growth can also occur in PE materials which are able to form craze zones at the tip of a propagating crack. Such intermittent crack growth can

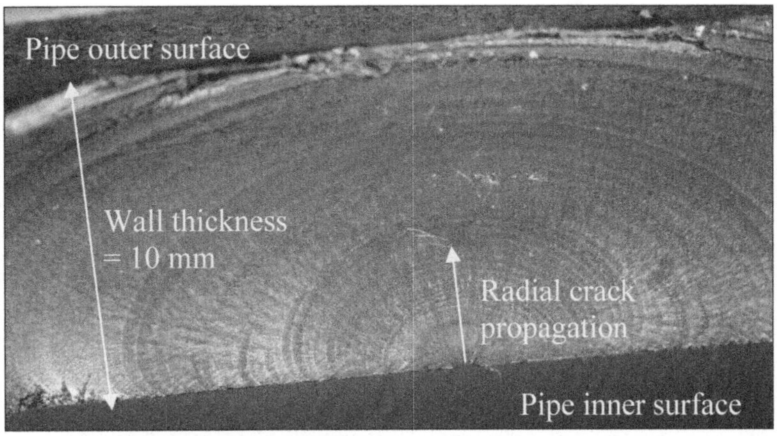

Figure 3.5 Fracture surface: Longitudinal fracture in 50mm diameter PE pipe. Installed in South Australia in 1994, failed 2004.

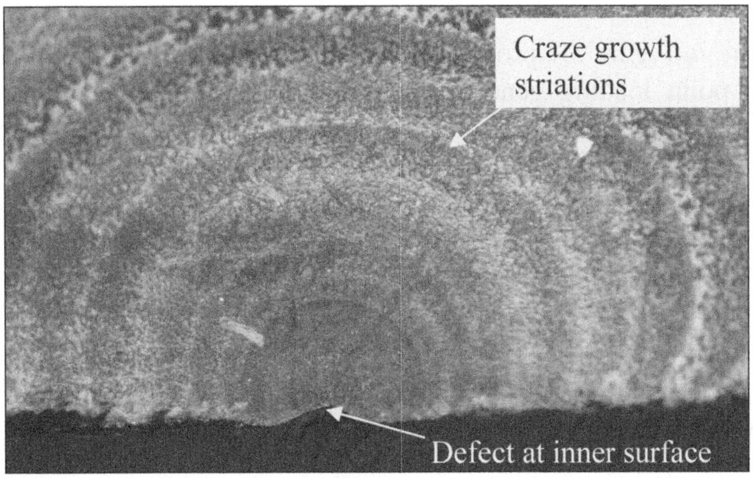

Figure 3.6 Fracture surface: Defect at pipe inner surface and striations showing craze growth and breakdown

occur under static load conditions and corresponds to the delay during which a craze is formed and then broken down (Parsons et al. 2000). Further examination of the pipe inner surface shows evidence of craze formation and breakdown (Figure 3.6).

Closer examination reveals increased whitening and local ductile deformation in the striations, which indicates formation, extension and subsequent collapse of craze fibrils. A detailed description of this intermittent crack growth mechanism in PE materials is given elsewhere (Parsons et al. 2000). Figure 3.6 also shows the point of crack initiation at the pipe inner surface. A small defect (approximately 0.1mm deep and 0.2mm long) is clearly visible at the centre of the emanating crack growth striations.

The fracture surface from the adjacent longitudinal crack is shown in Figure 3.7. Whilst craze striations are again visible, they do not remain elliptical for the entire crack path through the pipe wall. Around the mid-point of the pipe wall, the striations flatten across their upper region. As shown, this corresponds with a point load indentation at the outer surface of the pipe. Whilst

Figure 3.7 Fracture surface: Evidence of point loading at outer surface of pipe and corresponding distortion of elliptical craze growth striations

the previous fracture in Figure 3.7 propagated under a uniform applied tensile hoop stress (from internal pressure), point loading generates additional localized bending stresses. This would accelerate crack growth and result in earlier failure, which seems reasonable since this crack was found under a repair clamp. The reaction of a material to localized stress intensifications is an important factor in determining its capability to resist failure under actual service conditions. The long-term strength that is predicted based on the extrapolation of stress-rupture data from fails to consider the substantial effect that localized stress intensifications may exert on the strength behavior of the tested material. These intensifications can result in response to point loading and the effect of "external" flaws from inadequate thermal fusion, scratches and gouges. This may suggest that material field performance potential needs to be evaluated by additional measures to those which are solely based on the stress-rupture performance of carefully prepared pipe subjected to hydrostatic pressure only.

While the photographs in Figures 3.4 to 3.7 show how the SCG process occurs, additional information such as the material manufacture date, material density and pressure rating may provide some further insight into its likely level of SCG resistance. Unfortunately, no extra information to determine the nature of the PE material was available from water authority databases.

SLOW CRACK GROWTH MECHANISM IN PE MATERIALS

As stated above, vulnerability to slow crack growth failure is an important factor that influences the potential service life of a PE pipeline. Previous research into the behavior of PE gas pipe materials has increased understanding of this potential failure mode and is also applicable to PE water pipelines.

As with many pipe grade polymers, slow crack growth failures in PE pipes initiate from defects in the pipe wall. In addition to external damage, Barker, Bowman, and Bevis (1983) propose that crack initiation can occur from:

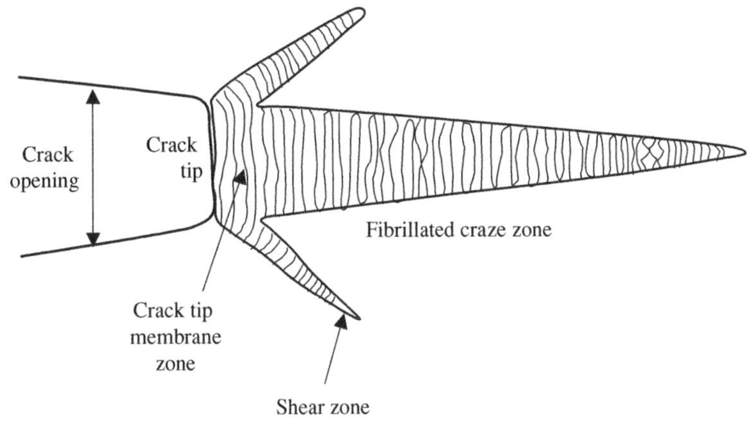

Figure 3.8 Schematic illustration of crack tip craze formed during slow crack growth in PE materials

- *Particles associated with base polymer production.* Several sources of crack initiation were revealed to be calcium-rich particles such as calcium stearate and/or particles of high molecular weight polymer or gels.
- *Particles associated with pipe manufacture.* Elemental analysis indicated that these particles are iron-based (316 stainless steel for example) and originated from wear and damage to plastics processing equipment.
- *Voids and cavities in thick walled pipes.* The origin of voiding was associated with two effects. Firstly, during production, rapid cooling from the outside surface can cause considerable residual stresses to be frozen into the pipe wall. The general distribution of these stresses is an outer region of compression balanced by an inner region of tension at the pipe inside surface, which may give rise to voiding. Secondly, polyethylene materials shrink substantially when solidifying. If the pipe outer surface is cooled first, then material at the pipe inner surface remains at a higher temperature and hence, a lower modulus. Consequently, this lower modulus material cannot deform the rapidly cooled higher modulus material to compensate for its shrinkage. Voiding can then occur.

After crack initiation, the relatively high slow crack growth resistance of newer PE pipe grade materials is conferred by a process zone, which is formed at crack initiation and develops during crack growth. This fibrillated zone or "craze" extends until separation and absorbs energy; increased energy absorption results in a higher resistance to slow crack growth.

As shown in Figure 3.8, a craze zone is comprised of fibrils of material drawn from the bulk as the crack tip opens. At the root of the craze zone is a plastically deformed "membrane" that separates the crack tip from the fibrillated craze zone. Extending out approximately 45° from the crack tip membrane are secondary zones of shear deformation. The time-dependent deformation (and ultimate failure) of the membrane and craze zones governs the slow crack growth and brittle fracture resistance of PE materials. Further evidence of the interaction between the membrane and craze zone is shown by periodic striations on the fracture surface of PE pipe failures (Sehanobish et al. 1985). This is thought to indicate a "stepwise" crack propagation mechanism in which the craze and membrane sequentially grow and fail, resulting in intermittent crack

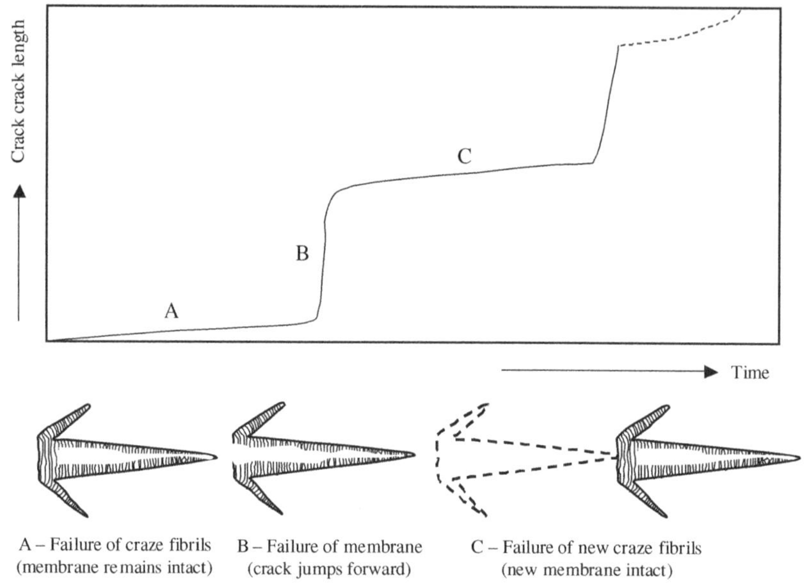

Figure 3.9 Schematic illustration of deformation process in stepwise crack propagation in PE

propagation. An example of fracture surface craze striations was shown previously in Figure 3.6. Stepwise crack growth in polymers is a consequence of their ability to relieve the stress concentration at a propagating crack by forming a quasi-stationary process zone (the craze) at the crack tip (Shah et al. 1998). Several comprehensive studies of stepwise crack propagation in PE have been conducted (Parsons et al. 2000, Shah et al. 1998). By observing the crack tip damage zone during long term tests, Parsons et al. (2000) described the deformation processes that occur in notched PE specimens.

As shown in the schematic diagram in Figure 3.9, the first plateau on a displacement/time record corresponds to deformation and separation of the craze fibrils. However, the membrane at the crack tip remains intact during this process. At some point, the membrane also ruptures and the crack tip advances through the previously ruptured craze zone. This produces the sharp increase in displacement shown at point B in the figure. The process then repeats itself with a second membrane and craze zone. Again, the craze fibrils deform and fail whilst the membrane is intact. Membrane failure then produces a second crack jump to point C on the displacement/time record. During stretching and crack opening, Shah et al. (1998) propose that the shear deformation zones at the crack tip delay the rupture of the membrane. After the main craze fibrils rupture, the stress originally supported by these fibrils is redistributed back to the membrane. It seems likely that deformation in the membrane is accompanied by further deformation in the adjacent shear zones. In a cold-drawing process, the adjacent shear zones supply material to the membrane allowing it to extend, delaying its ultimate rupture and the subsequent crack step jump in crack length.

The Influence of PE Material Microstructure on SCG Resistance

The formation and separation of a crack tip craze under static load is inherently related to material microstructure, which can vary from one PE grade to another. As discussed in chapter 1, PE is a semi-crystalline polymer that (upon cooling from the melt), forms inter-connected crystalline blocks (or lamellae) within a disordered, amorphous phase of molecular entanglements.

As described by Lin and Argon (1994), the crack growth resistance of pipe grade PE is related to the *tie molecules* that link these individual lamellae together in the crystalline phase. Upon loading, tie molecules stretch as they take up the applied stress and form taut links between crystalline lamellae. Crystals shear when the local stress generated by the taut molecular link induces crystallographic slip in the chain direction at the point of entry into the crystal (Lin and Argon, 1994). This leads to local yielding, breakup of crystals into smaller blocks and micro-void formation in the amorphous phase. As these micro-voids expand to the macro scale, the material between voids is drawn into highly oriented fibrils, which constitute the crack tip damage zone, or craze.

Under static loading, ultimate failure of craze fibrils between voids is associated with molecular chain disentanglement and pullout from crystalline lamellae (Shah et al. 1998). The structure and connectivity between crystalline and amorphous phases depends on the molecular weight and comonomer side chain distributions. Since these molecular characteristics differ between PE materials, then differences in SCG resistance are expected. As the molecular weight of PE materials increase, more tie molecules are formed during cooling from the melt, and consequently, more time is required for the molecular pullout process associated with fibril failure to occur. To verify this, Shah et al. (1998) demonstrated the increased stepwise crack growth resistance of PE homopolymers of high molecular weight compared to those of lower molecular weight. Introducing comonomer side branches (butyl branches for example) to the main PE molecular chain also increases the time required for chain disentanglement to occur. Short side branches act as protrusions along the chain and are not incorporated into crystalline lamellae. Consequently, these side branches "anchor" molecular chains between crystalline lamellae, again significantly delaying chain pullout and craze fibril failure (Lin and Argon, 1994).

The concept of tie molecules can be used to qualitatively explain the difference in crack growth resistance of new generation bimodal PE materials compared to conventional unimodal PE previously used for pipe applications. In unimodal PE, the chain branch content decreases with increasing molecular weight. Therefore, the anchoring potential is highest for low molecular weight, (short) PE molecules. However, these short molecules are least likely to be incorporated into several crystalline lamellae, which is the requirement to form an inter-crystalline tie molecule. Whilst increasing the comonomer branch density on these short PE chains would increase the density of inter-crystalline tie molecules, it would also reduce crystallinity and yield strength (Lin and Argon, 1994). In contrast, new generation bimodal PE materials have a wider molecular weight distribution and their comonomer side branches re-distributed to the higher molecular weight (long) PE molecules (Figure 1.1). The lower molecular weight fraction of these materials is essentially homopolymer, which increases crystallinity. The longer molecules participate in an increased number of crystalline lamellae but are anchored more frequently by comonomer side chains, which produce a greater number of intercrystalline links (Shah et al. 1998). Consequently, much longer times are required for chain disentanglement and pullout to occur, leading to failure of craze fibrils. Consequently, the latest generations of PE materials are highly resistant to slow crack growth, and the extended test times required to assess these materials are often impractical (Chan and Williams, 1981).

CONCLUDING REMARKS

Anecdotal evidence from the UK suggests that if surface preparation and installation are poor, electrofusion fitting failures can occur in PE pipe networks. Photographs from field failures indicate that a common failure mode is the expansion of moisture between EWF coupling and pipe outer surface. Under high temperature, moisture boils to steam and expands resulting in large cavities at the fitting/pipe interface. Alternatively, failure to properly clean and scrape the pipe surface to be joined can result in contamination of the joint and early failure. It should be noted however, that these failure modes are avoidable and with proper preparation and installation procedures, reliable, leak free heat fusion joints are obtained. The good performance record that has been achieved in gas distribution applications is in part due to the attention that the regulatory codes give to these factors (Mruk, 2006).

Depending on the PE material, slow crack growth failures can occur in PE pipes in service. Examination of fracture surfaces from these failures reveal elliptical striations which indicate the sequential formation and breakdown of a craze zone at the crack tip. The craze zone is a region of fibrillated material that forms at the crack tip, extends normal to the crack propagation direction and ultimately fails by fibril rupture. New generation PE materials can form large crack tip crazes with fibrils that resist extension and rupture and consequently exhibit increased resistance to slow crack growth failure. This resistance to craze growth and failure (and hence slow crack growth) depends on the density of intercrystalline tie molecules (links) that exist between individual crystalline lamellae in a PE material. The time-dependent failure of craze fibrils requires these tie molecules to be pulled free from their anchor points in crystalline lamellae. It is generally accepted that the density of tie molecules formed in a PE pipe is related to the molecular weight and short chain branch (SCB) distribution of the PE material. Unimodal PE materials usually exhibit a SCB distribution that is concentrated towards the PE chains of low molecular weight. Therefore, the anchoring potential is highest for low molecular weight, (short) PE molecules in these materials. However, due to their length, these short molecules are least likely to be incorporated into several crystalline lamellae, which is the requirement to form an intercrystalline tie molecule. In contrast, new generation bimodal PE materials have a wider molecular weight distribution and their SCB are re-distributed to the higher molecular weight (long) PE molecules. The lower molecular weight fraction of these materials is essentially homopolymer, which increases crystallinity. The longer molecules participate in an increased number of crystalline lamellae and are anchored more frequently by comonomer side chains, which produce a greater number of intercrystalline links. Consequently, much longer times are required for chain disentanglement and pullout to occur, leading to failure of craze fibrils. Consequently, the latest generations of PE materials are highly resistant to slow crack growth, and the extended test times required to assess these materials are often impractical.

CHAPTER 4
REVIEW OF PERFORMANCE STANDARDS FOR PE PIPES

INTRODUCTION

The term polyethylene (PE) when referred to PE pipes and fittings actually covers a broad series of resins with different, molecular weights, crystallinity, stabilizers etc, which are combined with colorants, ultraviolet inhibitors, stabilizers, antioxidants, etc, to form PE piping compounds. Polyethylene is unique amongst polymers commonly used for pipe materials in that there is continuing development of resin grades with higher strengths, as highlighted by the common resin classifications used in the majority of the world of PE63, PE80 and PE100, where the designated number indicates the long term hydrostatic strength in MPa. Additionally newer PE resins are characterized by improved characteristics in areas such as slow crack growth resistance. Whilst the resin structures in these grades may be varied and have significantly different properties, there are seven key performance criteria that ensure that the long term performance of a polyethylene pipe or fitting will be satisfactory. These are:

- Long term hydrostatic strength
- Fatigue resistance
- Resistance to slow crack growth (SCG) and rapid crack propagation (RCP)
- Adequate dispersion of additives
- Adequate stabilizer levels to ensure long term resistance to thermal loads and UV radiation
- Effect on water
- Quality of installation (including bedding, jointing, etc.)

The assessment of pipeline materials to these key criteria varies from country to country with standardization in the United States being covered by AWWA and ASTM. Internationally the ISO has developed a range of standards for PE pipes and fittings which are normally fully adopted or modified to meet local needs, as in the case of the Australian Standards. The purpose of this review is to highlight and review international standards which have been written for the standardization of PE pipes and fittings identifying and outlining the specifications and requirements for polyethylene pipe manufacture.

MATERIAL SPECIFICATIONS

Polyethylene material specifications are outlined in a number of standards; in general all pipes which are manufactured from polyethylene must contain antioxidants, UV stabilizers and pigments which are necessary for their manufacture into pipes or fittings. Physical material properties such as slow crack growth resistance, density, tensile resistance and flexural modulus are found to provide an indication as to how well these materials will behave and perform in service. Table 4.1, outlines the polyethylene pipe manufacture standards and their relevant material classification systems for detailing specific material requirements.

Under current practices the classification of PE pipe materials and the definition of their minimum property requirements is split between a material classification standard and a pipe

Table 4.1
PE material classifications for polyethylene standards

Pipe Standard	Pipe material	PE material classification	Relevant Standard
AS/NZS 4130	PE	PE80 / PE100	AS/NZS 4131
ISO 4427	PE	PE32, PE40, PE63, PE80, PE100	
ASTM D2239	PE	Cell classification	ASTM D3350
ASTM D2447	PE	Cell classification	ASTM D3350
ASTM D3035	PE	Cell classification	ASTM D3350
ASTM F714	PE	Cell classification	ASTM D3350
AWWA C901 - 02	PE	Cell classification	ASTM D3350
AWWA C906 - 99	PE	Cell classification	ASTM D3350

NOTE: ASTM 2239, covers PE pipes made to standard inside diameters and as such is seldom used in municipal applications. ASTM F714 are made to outside dimensions and thus more likely to be used in water distribution applications, however, for such application s water utilities generally specify the listed AWWA standards.

specification standard as shown in Table 4.1 which outlines the main standards referenced for the manufacture of polyethylene pipe and their respective material classification standard. However the split of specifications is determined by the standards writing body, for example the ISO and Australian standard do not provide any material classification systems such as the cell classification system utilized in the United States in accordance with ASTM D3350. However the ISO 4427 and Australian Standards AS/NZS 4131 do specify a number of critical minimum material requirements that must be meet such as Melt Flow Rate, Thermal stability, carbon black content (for black pipes), Dispersion of additives, effect on water, volatiles content, rapid crack propagation and weathering resistance.

Both the ASTM and AWWA standards commonly use both a standard PE code and a cell classification system as identified in accordance with ASTM D3350. This cell classification system classifies materials in accordance with the test values obtained from the following primary material properties, density, melt index, flexural modulus, Slow Crack Growth Resistance, tensile strength at yield, environmental stress crack resistance and the hydrostatic design basis at 23°C. Physical properties such as ultraviolet stabilizer levels, thermal stability and elongation at break are also included.

The ASTM standards D2447 and D2239 are confusing in that they call up specific material specifications of Class C, Grade P14, P23, P24, P33 and P34 from ASTM D1248 (Polyethylene Plastics Extrusion Materials for Wire and Cable) or they say that the PE plastics can also be described in accordance with the appropriate cell classification as defined in ASTM D3350. Assessment of ASTM D1248 could not find the relevant grades referred to in the pipe specifications and there does not appear to be any relationship between the material specifications and the cell codes. However, it has recently been agreed that D3350 should be exclusively referenced for pipe and fitting applications and the standards will be upgraded accordingly. ASTM F714 was a little clearer in that it specified that resin compounds shall meet the general physical requirements listed in D3350, without giving an actual cell code needed for their materials. In contrast, the AWWA standards C901 and C906 are clear in that they require the materials used in pressure pipe to contain the following minimum material classifications as shown in Table 4.2.

Table 4.2
Minimum material requirements in AWWA standard

Standard PE code	Cell classification
PE2406	PE213323
PE3406	PE324433
PE3408	PE334434

NOTE: The basis for assigning standard PE codes in D3350 is continually evolving as more information is obtained on material properties and their implications on performance, for example PE3408 may now carry the designation PE4708 and a PE4708 may be designated as a PE4710.

Table 4.3
Comparison of material properties from polyethylene pipe standards

Material property	AWWA C901 & C906	ISO 4427-1	AS/NZS 4130	ASTM F714 D3035*
Density (g/cc)	> 0.940–0.955	≥930	—	> 0.940–0.955
Melt Index	<0.4 to 0.15	0.2–1.4	—	<0.4 to 0.15
Flexural Modulus (MPa)	552–< 758		—	552–< 758
Tensile Strength (MPa)	21–< 24		—	21–< 24
Slow Crack Growth	3	No Failures	500 hours*	3
HDB at 20°C (MPa)	8.62	8 (ISO 4427-2)	8 (AS/NZS 4131)	8.6

* ASTM F714, D2447 and D2239 pipe standards allow for different material classifications such as PE2406 and PE3408. The material values shown in the table are for PE3406 material. Slow Crack Growth resistance results and HDB test results are difficult to compare because of the different test methods.

It should be noted here that recent changes to D3350 have resulted in changes to the code designations. This code—which ASTM defines as the "thermoplastic pipe materials designation code" defines stress rated PE piping materials based on the following three criteria:

- The letters (PE) identify that it is a polyethylene based composition
- The first two numbers originally identified the Type and Class in accordance with ASTM D1248, but they now have been revised so they identify the density and the slow crack grow resistance cells in accordance with the recently revised D3350
- The last two numbers identify the maximum recommended hydrostatic design stress (HDS) category for water, for 73.4°F. These last two numbers represent a materials HDS category in psi after dividing it by 100.

The individual minimum material property requirements are compared to that of the Australian standard in Table 4.3 for the equivalent of a PE80 material. In addition to the test requirements detailed by the cell classification additional tests are required for thermal stability and brittleness temperature.

It is difficult to compare standards with relation to material properties, for example the ISO 4427-1 and AS/NZS 4130 standards do not outline many of the specific material requirements for the resin that are specified in ASTM standards, but instead specifies performance requirements for factors such as consistency of melt flow, thermal stability, carbon black levels size and toluene extraction, dispersion, weathering resistance, volatiles level, short term resistance to internal pressure, slow crack growth resistance, effect on water and resistance to rapid crack propagation when manufactured into a pipe for PE100 materials. Referring to AS/NZS 4131 *polyethylene compounds for pressure pipes and fittings,* material properties such as those outlined in the ASTM standards including, density, tensile strength at yield, flexural modulus and melt index are not specified.

From the review conducted here it would appear that both the ASTM and AWWA standards for polyethylene pipe provide a detailed material specification when compared to that of the Australian standards AS/NZS 4130 and AS/NZS 4131 as well as the ISO standards on which they are based. The US standards clearly detail the specific material requirements required for compliance, through the use of the cell classification system. This method of classification provides a detailed grading system based on material properties which outlines the specific material requirements that a material must exhibit. The ASTM material classification system allows any new requirement that test and experience have demonstrated as significant to be included. For example, soon after method ASTM D1473 (the accelerated method for measuring resistance to SCG) was first issued, ASTM D3350 was modified to include requirements that are based on the results obtained by means of this method (Mruk, 2006). In this context the ASTM cell system may be overly prescriptive as it specifies a significant number of requirements that are effectively retested in some of the latter pipe tests. As detailed in the introduction, the performance of PE pipes is determined by seven criteria discussed below and an alternative may be to relax the cell criteria by removing some of the non-essential requirements and ensure that adequate performance is addressed by relevant performance standards in the pipe standards (This would have the benefit of aligning the USA closer to the strategy adopted in ISO standards). Each of these areas is described in more detail below. There are some test criteria that should be addressed at the compound level that seem to have been missed in the United States standards. For example, carbon black sizing and dispersion, weathering resistance etc. are issues that may influence pipe performance, but are essentially compound characteristics and should be addressed at that level. However, their inclusion could be a contentious issue in the United States as some manufactures claim from experience that factors such as carbon black sizing and dispersion are not critical factors with regards to performance, as long as carbon black is included at levels 2%. In this respect it should be noted that Australian experience has shown that carbon black quality can have a significant impact on performance in areas such as impact on water quality.

LONG-TERM HYDROSTATIC PRESSURE RESISTANCE

Hydrostatic pressure testing is used to establish a hydrostatic design basis (normally at 50 years) which is then divided by a safety or design factor to establish a hydrostatic design stress from which the working pressure for each pipe size and pressure rating is calculated, using the ISO thin walled formulae. Hydrostatic pressure testing has typically involved testing a number of individual pipe samples by applying varying levels of internal hoop stress or internal pressure to cause failures along a linear relationship ranging in failure durations between 10 and 10,000 hours. Under ISO 9080 *Plastics Pipes and Ducting Systems—Determination of the Long-Term Hydrostatic Strength of thermoplastics Materials in Pipe Form by Extrapolation,* the hydrostatic strength is obtained by testing a minimum of 30 specimens spread regularly over 5 test intervals (tested at 20°C) and then extrapolating the test results to a period of 50 years using regression techniques to obtain the 97.5% lower confidence limit of the average value of its predicted hydrostatic strength. This 97.5% lower confidence limit (labeled a Minimum Required Strength or MRS) is then divided by a design coefficient, which varies between 1.25 and 3.2 to give a maximum allowable hydrostatic design stress. Based on this figure the pipe material is then classified into one of a number of recognized standard strength categories, i.e., PE80, PE100.

In comparison to ISO 9080, ASTM D 2837, *Obtaining Hydrostatic Design Basis for Thermoplastic Pipe,* uses a minimum requirement of 18 test points between 10 and 10,000 hours and then extrapolates these points using linear regression to the intercept with 100,000 hours. The value at the 100,000 hour intercept is called the long term hydrostatic strength (LTHS). Based on the LTHS obtained, the Hydrostatic Design Basis (HDB) category of the tested material is assigned. The Hydrostatic Design Stress (HDS) is determined by multiplying the HDB by a design factor, which is approximately 0.5 in the case of C901 pipes.

The essential similarities and differences between the ISO 9080 and ASTM D 2837 are as follows (Mruk, 2006):

- Both methods evaluate "new" pipe that is only subjected to internal hydrostatic pressure. Effects of flaws and of stress concentrations that are induced by external factors are not considered.
- In both methods, hydrostatic pressure is developed using water as the medium.
- Both methods are empirical in nature. The nature of the failure is not a factor in the analysis and interpretation of the stress-rupture data.
- Method D2837 requires that the obtained data be obtained through a test period of not less than 10,000 hours and that, when plotted on log-log coordinates, the data approximately follow a straight line behavior. If the data obtained on a PE pipe material meet the above straight line requirement through the first 10,000 hours the resultant straight line may then be extrapolated to determine the long-term strength at a longer time.

For older PE compounds a rapid decrease of strength with extended loading durations can occur which leads to a second straight line relationship between pressure and time. This can be associated with a shift into a brittle-like failure mechanism which, due to localized stress concentrations that are generated by inherent flaws, causes small cracks to be incubated which then subsequently slowly grow through the pipe wall resulting in failure. This mechanism is labelled slow crack growth (SCG). This shift in failure mechanism is greatly accelerated by increasing the temperature at which stress-rupture testing is conducted. To account for this non linearity in the test points which is associated with the different failure mechanism associated with slow crack growth (SCG), standards provide additional requirements to determine linearity. US standards account for this issue in a different manner to international standards such as ISO 9080:

- For method D2837, where it is uncertain that linearity exists in the data, it is not possible to assign a hydrostatic design basis to the material. In contrast method ISO 9080 has been designed to accept data that show a departure from a straight line in its log-log behavior.
- To ensure non linearity does not occur ASTM D2837 requires an additional requirement must be satisfied: The results of a "rate process" analysis (Palermo and De Blieu, 1985) of elevated temperature stress-rupture results must "validate" the

assumption of continuing straight line behavior through at least the 100,000 hour intercept. Relatively recently, an option has been added—called "substantiation"—which is designed to validate this assumption through at least the 50-year intercept (Mruk, 2006).

- Method ISO 9080 also requires elevated temperature stress-rupture data be collected but, it uses these data to forecast a long term strength independent of whether this strength lies along the straight line that is defined by the earlier data or, in a down-turn region that may be predicted by an analysis of the elevated temperature test results. To accomplish this, method 9080 requires considerably more elevated temperature stress-rupture data which then are analyzed against certain mathematical models. The model that results in the best fit is used to predict a PE's longer-term ambient temperature strength.

The apparent advantage of the ISO method is that it forecasts the long-term hydrostatic strength of a material even though that strength may fall in the down-turn region that is governed by the brittle-like failure mechanism associated with slow crack growth. In contrast, one potential advantage of the ASTM method is that it simply excludes materials that exhibit a downturn in their stress-rupture properties (Mruk, 2006). It should be noted however, that for modern PE's, which exhibit very high resistance to SCG, ambient temperature stress-rupture behavior follows a straight line through at least the 50-year intercept. It has been demonstrated that for these materials, methods ASTM D2837 and ISO 9080 forecast essentially the same long-term strength (Boros, 1995).

A difference in the hydrostatic design stress that results under the ASTM and ISO protocols is mostly due to the method by which the estimated long-term strengths are placed in standard strength categories and by the choice of the strength reduction factor by which a strength category is reduced into a hydrostatic design stress (Mruk, 2006). A comparison of how HDS values for the design temperature are established using ASTM D 2837 and ISO 9080 methods is summarized in Table 4.4 (Mruk, 2006).

As shown in Table 4.4, there is one important difference between the two design protocols. The North American protocol for the design of PE pipe intended for the conveyance of water uses an **overall** strength reduction factor, whereas under the ISO protocol only a material factor is used—the choice of the overall factor is left up to the local authority.

Table 4.5 summarizes the required Hydrostatic Design Stresses from each of the ASTM, AWWA and ISO standards for polyethylene pipe manufacture. Where given, safety and surge allowances have also been identified.

It is evident from an inspection of Table 4.5 that higher design stresses are realized under the ISO and consequently standards such as the Australian/New Zealand standards, which are based on the ISO. Part of the reason for this is the use of a lower design temperature of 20°C (68°F) at which PE is about 5% stronger than at 23°C (Mruk, 2006). Another reason for this difference is that a smaller strength reduction factor is used. As stated above, the ISO standard establishes a pipe's HDS by only applying the *basic* coefficient. In contrast, US standards establish the HDS based on a more conservative *overall* coefficient. This is thought to be more appropriate for pipes that are to be part of a system that includes fittings and joints, components that can generate localized stress concentrations that are not present during the stress-rupture evaluation of pipe (Mruk, 2006).

Another difference that applies to the PE100 material is that under the ISO long-term strength categorization system, certain third generation materials have been able to enter the MRS

Table 4.4
Establishing Hydrostatic Design Stress (HDS) values from ASTM D 2837 and ISO 9080 methods

	Current procedure	
Action	Per ASTM/AWWA protocol	Per ISO protocol
Design temperature	73.4°F (23°C)	68°F (20°C)
Determination of the long-term strength (LTHS)	The average value of the long-term hydrostatic strength (LTHS) that is forecasted at the 100,000 hour intercept. For this purpose, the LCL of the LTHS may not be less than 85% of the predicted average value of the LTHS as per method D2837.	The 97.5% LCL of the average strength that is forecasted at the 50-year intercept. Per Method ISO 9080.
Reduction of the LTHS into standard hydrostatic strength category	Using a 10-step preferred number series (e.g., 1,000 psi, 1,250 psi, 1,600 psi, etc.). The result is called the hydrostatic design basis (HDB).	Using a 10-step preferred number series (e.g., 6.3 MPa, 8.0 MPa, 10.0 MPa, etc). The result is called the minimum required strength (MRS).
Reduction of the standard strength category into a hydrostatic design stress (HDS)	Multiply the HDB by an overall design *factor* of 0.50 which has been established in consideration of all the variables that may affect the long-term strength in a PE piping installation and, also in consideration for the achieving of a service life very significantly longer than the 100,000 hour intercept at which the LTHS was determined.	Divide the MRS by a basic *design coefficient* of 1.25. (ISO calls this 1.25 value as a material coefficient and it recommends consideration of a possible add-on coefficient for pipe that is part of a piping system that is exposed to conditions that are significantly different from those represented by the method used for the determining of the pipe material's MRS).

10 category. Although these materials display only a relatively small gain in strength over materials that are classified as MRS 8.0, the gain is sufficient for them to be classified as PE100 and thereby, be considered as being 25% stronger than the PE80 materials. In contrast, under the ASTM categorization system, these newer materials stay within the same 1,600 psi HDB category. Because the ISO categorization system awards these third generation materials with a higher strength designation, some commercial interests in North America have favored the adoption of the ISO system for the pressure rating of PE pipe. The response by US standard writing groups has been to recognize materials that offer superior SCG resistance toughness—not only the higher density PE100+ materials but also medium density materials—by determining that for such materials the design factor may very safely be increased be increased from 0.50 to 0.63 (Mruk, 2006). This is equivalent to the raising of the HDS for these materials by a factor of 25%. This recommendation has been accepted and is in the process of being implemented into all applicable standards. These PE materials must comply with a number of additional material properties, such as high resistance to slow crack growth, no evidence of a knee in the regression data and lower levels of data scatter. This has resulted in two new material categories PE3710 and PE4710 for which the

Table 4.5
Hydrostatic design stresses for polyethylene pipe

Standard	Material	HDS (MPa)	Design factor	Surge allowance
ASTM D2447	PE1404	2.76	—	—
	PE2305	3.45	—	—
	PE2306	4.34	—	—
	PE2406	4.34	—	—
	PE3306	4.34	—	—
	PE3406	4.34	—	—
ASTM D2239	PE1404	2.76	—	—
	PE2305	3.45	—	—
	PE2306	4.34	—	—
	PE2406	4.34	—	—
	PE3306	4.34	—	—
	PE3406	4.34	—	—
	PE3408	5.52	—	—
AWWA C906	PE2406	4.34	0.5	*
	PE3406	4.34	0.5	*
	PE3408	5.52	0.5	*
AWWA C901	PE2406	4.34	0.5	*
	PE3406	4.34	0.5	*
	PE3408	5.52	0.5	*
ISO 4427†	PE80	6.3†	1.25†	—
	PE100	8.0†	1.25†	—

* The design factor of 0.5 used for the AWWA standards allows for limited surge pressure effects which may occur.
† In the ISO 4427 standard, hydrostatic design stress (HDS) has been specified for series 1 PE80 and PE100 as being 6.3 and 8.0 MPa respectively taking into account a basic service design coefficient of 1.25.

HDS is 1,000 psi (6.9 MPa). If the temperature difference between ISO and ASTM/AWWA standards are taken into account, the HDS at 20°C for these materials is 7.2 MPa, which can be compared to the ISO value of 8.0 MPa. However at this stage there is no material classification equivalent to the ISO PE100 class with a HDS of 10.0 MPa.

A number of short term testing parameters are also outlined in both the ASTM and Australian standards for polyethylene pipe and are discussed in the following sections.

Table 4.6
80°C performance requirements for internal pressure resistance in accordance with AS/NZS 4130

Compound classification	Applied stress (MPa)	Min. time to rupture (hours)
PE80	4.6	165
	4.0	1,000
PE100	5.4	165
	5.0	1,000

SHORT-TERM HYDROSTATIC PRESSURE RESISTANCE

Whilst hydrostatic pressure testing on pipe specimens extruded from PE pipe compounds is used to ensure the long term hydrostatic design stress, this testing methodology is also used in a number of standards to ensure consistency in the manufacturing process. For example Australian Standard AS/NZS 4130, specifies a number of hydrostatic pressure performance requirements at 80°C as noted in Table 4.6 for both PE80 and PE100 materials.

It should be noted that the test stresses that are specified for the 80°C sustained pressure test are in a region that has been determined not to result in failure by the ductile mode—the intent is to test the material where failure, if it occurs, will be by the brittle-like failure mode. Sustained pressure testing at these stresses at 80°C is an accelerated means for the evaluating of a material's ambient temperature resistance to failure by the SCG mode. If a premature failure is noted, that event is recognized as an indicator that the material fails to offer the expected minimum degree of resistance to SCG.

Similarly to AS 4130 requirements, ASTM D2239, D2447, F714, D3035 and AWWA C901 and C906, require sustained pressure testing, but at differing temperature, pressures and times, compared with AS4130 which stipulates 80°C with an applied stress of either 4.0 or 4.6 MPa for 170 and 1,000 hours respectively. Additionally there are a number of short term quick burst test. For example AWWA C901 tests at 2,500 psi hoop stress and requires that the failure be ductile in nature. Whilst the relative benefits of the different short burst methods contained in the methods could be debated, it would appear that there is a need for a short term pressure test at ambient temperatures to ensure that ductility is achieved at a point close to the requirements of the hydrostatic test line as well as a consistent sustained pressure test at elevated temperatures and longer time frames to ensure that slow crack resistance is not evident. There are currently studies underway to develop a consistent test procedure to evaluate the SCG resistance properties in a relatively short time frame.

With regard to the first point, pipe manufacturing plays a critical role in the ability to meet long-term pressure requirements. Continuous tests such as the short burst tests detailed above are normally specified for the manufacturer to ensure that the ability to meet the HDB is maintained. However these tests are prescriptive and do not address the changing requirements of material selection or the pipe manufacturing process. What these tests need to ensure is that the manufacturing process is of a high enough quality so that the manufactured pipe will continue to meet the long-term hydrostatic performance requirements. This can be achieved by selecting samples on an ongoing basis and testing them to see if they are within the same statistical variation as those that were used to develop the hydrostatic regression line. Unfortunately, this requires the manufacturer to have undertaken sufficient tests to develop this regression line in the first place. This methodology is being adopted throughout the range of Australian standards for pipeline materials, and has been proposed to a range of ISO committees to cope with the varying compounds and manufacturing techniques now associated with PVC and PE pipe manufacture. This is generally proposed as a one-hour test with a hoop stress (S) value equal to the 99.5% LCL value at one hour, determined from the 50-year linear regression analysis for the appropriate material, as detailed ISO 9080. By testing according to this requirement, the variations associated with material selection and the manufacturing processes are eliminated, and the test ensures that the manufacturer is assessing the ongoing samples to a satisfactory performance standard.

SURGE AND FATIGUE

Surge or water hammer as it is more commonly known is caused by a sudden change in the water velocity along a pipeline. These sudden changes in water velocity can cause the internal pipe pressure to reach a level that exceeds that of the class of the pipe in question. Common sources of surge or variations in pressure can be caused by pump start up or opening and closing of valves. If surge events are found to occur at a frequent rate during the pipes lifetime, fatigue to the material can result in premature failure of the pipeline. It is the author's opinion that design for fatigue and surge is a specialist area and allowance for such events should not automatically be included in the design of pipelines which are normally installed in systems subjected to pressure variations associated with diurnal water demands.

Where surge events are likely to occur and are specifically being designed into the pipeline, safety factors or allowances need to be factored into the maximum working pressure which a pipe is allowed to operate at. Whilst it may not have been the initial intention of these standards, detailed examination of both of the AWWA standards C901 and C906 gives the impression of an allowance for surge being incorporated into the pressure rating of the pipe, but not the components, with an assumption that the total number of surge events that the piping experiences over its intended service life will not exceed 1 million. This is compounded by the use of a design factor of 0.5, compared to the ISO multiplier of 0.8, which effectively increases the pressure class by 38% compared to ISO, with the corresponding wall thickness penalties and increased costs to the water utilities. However, it should be noted that this comparison does not include the different test temperatures, the use of mean vs. LCL extrapolated values and the different time intercepts, which will narrow the margin. However, regardless of these issues, the current AWWA pipes are significantly more conservative in their design than the equivalent ISO pipes. As noted earlier, AWWA standards will be changed to allow materials that exhibit a high resistance to slow crack growth to utilize a design factor of 0.63, an imitative which will considerably narrow the difference.

As discussed by Burn et al. (2005), the practice of allowing for surge is questionable and for most pipelines operating under static pressure it is not needed, with scientific evidence suggesting that polymeric pipes can cope with occasional over-pressurization satisfactorily as under these circumstances the pipe becomes stiffer and the pipe can sustain the associated additional load (Marshall and Brogden 1997). As discussed by Burn et al. (2005), the demonstrated ability of plastics pipes to accommodate short-term pressure surges can be utilized to its advantage, by allowing some over-pressurization to the pipe. However for fittings, generally the safest approach would be to design so that the peak pressure for the fitting does not exceed the field test pressure of the pipe, which is normally 1.25 times the maximum operating pressure.

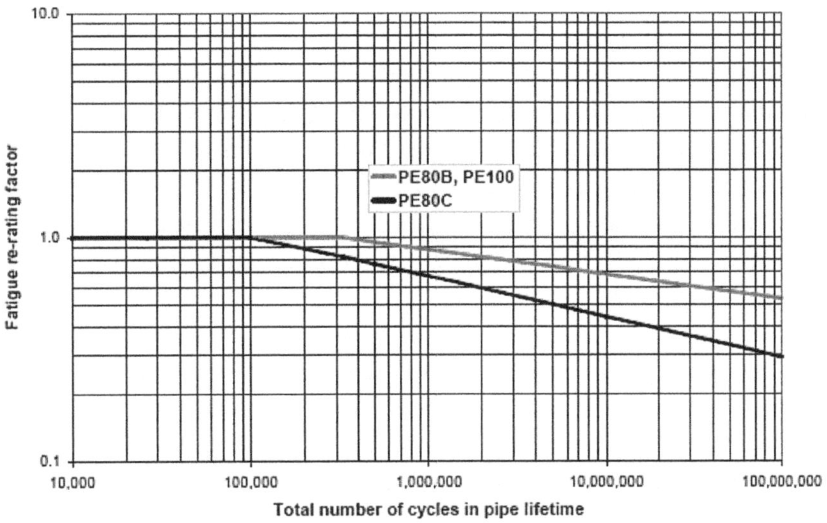

Figure 4.1 Suggested fatigue de-rating factors (see appendix B) (adapted from PIPA, 2002a)

If occasional pressure is irrelevant and the design issue is actually cyclic pressures, then the pipeline should be designed specifically for fatigue conditions and the application of a surge allowance is a simplistic approach. Fatigue in PE is a slow crack growth or craze extension mechanism which is controlled by a number of factors including the presence of stress concentrators, surge amplitude, frequency and mean pressure and pipes subjected to such conditions should be specifically designed for these conditions. It would be possible to use the current methodology detailed in AWWA standards where surge is broken down into three scenarios being surge pressure, recurring surge pressure and occasional surge pressure to design for surge in a consistent approach. However, whilst this methodology has merit, it does not allow for de-rating the pipes for different surge frequencies. An alternative approach is detailed in appendix B. In this methodology (PIPA 2002a) a de-rating factor is established as shown in Figure 4.1 by taking into account the surge amplitude, the surge frequency and the required pipe lifetime. In addition to derating factors for pipelines, a methodology for derating of fittings has also been established (PIPA 2002b).

RESISTANCE TO RAPID CRACK PROPAGATION

Rapid Crack Propagation is a phenomenon that can be associated with the use of thick walled pipes carrying high pressure gases operating in colder climates. In failures of this type, the high energy levels available in compressed gas can effectively drive a crack in a failed pipe many meters along its length from the point of initiation. As discussed in chapter 3, failures of this type in water pipelines are extremely rare and consequently, standardization for water applications has not required a performance requirement in this area. However, to design against possible RCP failures in pipes containing entrapped air, Australian/New Zealand Standards (AS/NZS 4131) has introduced a requirement in this area for resins that are classified as PE100 grade. ISO 4427-1, whilst stating that RCP needs to be taken into account for pipes greater than 32mm in wall thickness and providing a test method in Annex B, does not actually require assessment against this test method. Neither ASTM nor AWWA standards address the issue of RCP at this stage.

RESISTANCE TO SLOW CRACK GROWTH

Slow Crack Growth (SCG) is a failure mechanism that occurs in PE resin and it is the failure mechanism that causes a change in slope commonly referred to as the 'knee' in hydrostatic pressure testing. Following field failures from earlier grades of PE, manufacturers have put considerable effort into manufacturing grades of PE that are resistant to SCG, to ensure that satisfactory life is obtained in the field. To reflect these improved grades and to ensure high performance standards have introduced a range of tests to ensure that PE resins used for pipe manufacture have resistance to SCG.

The tests to ensure that SCG does not occur in PE pipes are split up into type tests on the resins and performance tests on the manufactured pipes. In the first category, ASTM 3350 specifies that resins shall either pass a requirement for Environmental Stress Crack Resistance according to ASTM D1693 or shall meet a requirement according to the PENT test of ASTM F1473. These tests are significantly different, ASTM D1693 subjects a small test specimen at an elevated temperature (50° or 100°C) under stress to a stress crazing agent, such as soap and measures the time to failure, whereas ASTM F1473 subjects a notched specimen to a constant load at an elevated temperature. Because of their cell classification most modern resins would be tested according to ASTM F1473. In comparison, ISO and Australian standards require conformance to ISO 13479. In the ISO 13479 test a full scale pipe specimen with four equidistant notches (around the circumference), notched through 20% of its wall thickness is subjected to hydrostatic pressure at 80°C for a period depending on the resin grade. As discussed by Beech, Ferguson and Clutton (2001), notched tensile specimens provide a more severe test than the likely service conditions for a pipe. In particular, they do not account for the crack tip blunting mechanisms which are likely to occur in circumstances of low notch tip constraint (i.e., surface notches in pipes). As a consequence, while notched tensile specimens are an excellent test to define the inherent SCG resistance of a pipe, they cannot be used in isolation to predict actual pipe performance. This is not to suggest that the ISO test can be used at this stage to predict actual field performance although further work is occurring in this area.

In the case of testing on manufactured pipes, ASTM D2239 and D2447 prescribe an environmental stress cracking (ESCR) test where specimens coated with soap (Igepal CO-630) are pressurized for 3 hours at 23°C. ASTM F714 and AWWA standards for PE pipes do not have a requirement for testing pipes for resistance to slow crack growth. Whilst ISO do not have any testing requirement in this area for manufactured pipes, Australian standards have a more rigorous requirement to ensure manufactured pipes will not fail by this mechanism and again call up the ISO 13479 standard. Because manufacturing can have a significant effect on compound quality in the final product, the adoption of a post manufacturing test in this area is necessary. Whilst the test method ASTM D2239 may detect issues with older type PE materials, it is recommended that their ability to discriminate poor manufacturing techniques in PE100 materials be investigated further. Alternatively a test such as that detailed in ISO 13479 could be considered to replace such test. While it could be argued that the sustained pressure, elevated temperature test would meet this requirement, it could also be argued that the addition of the notches as in ISO 13479 is more discriminatory on resins prone to SCG.

ADEQUATE DISPERSION OF ADDITIVES

Adequate dispersion of additives is an essential factor if extended service lifetimes are to be achieved, as inadequate dispersion of additives as well as providing sites for slow crack growth initiation, can also allow degradation from UV and heat in areas with low stabilizer levels. ISO standards have a performance requirement in this area for the compound specifying a dispersion level of ≤ grade 3 for both pigments and carbon black, it being assumed that if these two additives are well dispersed then other additives such as stabilizers will be too. Whilst ISO has no requirement at the manufactured pipe level, Australian/New Zealand standards (AS/NZS 4130) require adequate dispersion to be shown for striping or jacket compounds again to the equivalent of the ISO requirement. Standards in the USA do not have a test requirement in this area and both ASTM and AWWA standards for polyethylene piping should consider adopting such a test to ensure adequate dispersion of all additives in the manufactured pipes.

LONG-TERM RESISTANCE TO THERMAL LOADS AND UV RADIATION

The stability of PE resins under UV degradation and high temperature conditions depends on the stabilization system used, and the level of stabilizers incorporated. Common screening agents/stabilizer systems used in PE are Carbon Black and Hindered Amine Light Stabilizers (HALS) systems each of which protects the PE by a different mechanism. If adequate stabilizers are not incorporated, premature failure associated with stage III chemical failure can occur (where stage I is ductile failure and stage II is failure associated with slow crack growth). When PE is exposed to heat or UV radiation in the 290–400 nm band, a series of complex reactions occurs which cause the PE to degrade, especially in the presence of oxygen. The accepted theory is that irradiation excites the PE polymer chains, causing radicals to form, which may further chain-react with the PE causing bond cleavage, or crosslinking. The radicals also react with oxygen to initiate chain cleavage or crosslinking, and form the hydroperoxide and carbonyl groups found in degraded polymer. Consequently, stabilizers are needed to protect PE from heat degradation (thermolysis) as well as light degradation (photolysis). The level of stabilizers added is generally controlled by the thermal performance needs of extrusion, and little consideration is given to the long-term thermal performance requirements due to weathering, although for long-term performance at elevated temperatures this needs to be considered.

Best practice in this area details testing for both UV and heat stability. Heat stabilizer tests are normally carried out on both the compound (before manufacture) and the pipe after manufacture. This is especially important for thick walled pipes where inadequate cooling on the inside of the pipe can lead to utilization of all the heat stabilizers in this area and the formation of microcrazing on the inside surface of the pipe. In contrast, UV resistance tests generally only need to be carried out on the compound. ISO standards such as ISO 4427 specify an Oxygen Induction Time (OIT) to ensure that adequate heat stabilizers are incorporated into the resin and that these stabilizers have not been fully consumed during manufacture. For both compounds and manufactured pipes ISO 4427 specifies an OIT of 20 minutes at 200°C with the specimen taken from the inside of the pipe for the manufacturing test. Australian Standards place a higher requirement of the compound test requiring an OIT of 40 minutes at 200°C.

All USA standards place a heat stabilizer requirement on the base compound and call up ASTM 3350. However, this places a different performance requirement on the compound, requiring the induction temperature to be ≥220°C when a compound specimen is heated from a base of 150°C at 10°C/minute. AWWA place a similar requirement on manufactured pipes requiring the induction temperature for a specimen taken from the inside wall of a pipe to be ≥220°C. In comparison, ASTM standards D2239, D2447 and F714 do not define a test requirement for thermal stability for the manufactured pipe and the adoption of a test methodology in this area should be considered for manufactured pipes.

UV resistance is provided to PE pipes by the inclusion of either Carbon Black or a free radical scavenger such as HALS in the resin. At an international level, the weathering resistance of PE pipe compounds is deemed satisfactory if they contain 2.25±0.25% Carbon Black or if they contain enough stabilizer so that when exposed to at least 3.5 GJ/M^2 of Global radiation they pass a short term hydrostatic pressure test as well as have an elongation at break of 350%.

ASTM 3350 requires a minimum level of Carbon Black of 2% and in this respect, follows International Standards for black materials. For other colors there do not seem to be any performance requirements specified, apart from the generic requirement to contain sufficient stabilizers for 24 months of outdoor storage in AWWA C906. This omission appears to have been promulgated into the pipe manufacturing standards for both AWWA and ASTM where there are no weathering requirements, for PE pipes for colors apart from black, except for the requirement of AWWA C906 and the statement in ASTM standards of "Class B compounds shall have sufficient UV stabilizer to protect pipe from deleterious effects due to continuous outdoor exposure during storage and shipping" and in AWWA standards of "The material shall contain pigmentation and ultraviolet stabilizer to meet the requirements of a C, D or E classification of ASTM D3350, Sec. 6.2." It would appear that this is an oversight for non-black PE pipes in ASTM and AWWA standards that should be addressed.

Where there is potential for either long-term UV or heat degradation, for example where PE pipes are exposed or carry water at elevated temperatures (some southern areas of the USA for example), the PE pressure pipe standards should consider incorporating suitable performance requirements to ensure long-term durability. While this report does not make specific suggestions on a representative test requirement, there are several suggestions that can be considered in the attempt to reduce the number of potential failures related to chemical degradation in PE pipe applications, viz.:

- Protection from UV exposure and elevated temperatures during storage and transport should be specified in the appropriate installation codes, as for example in AWWA M23. While this current standard only specifies requirements for storage less than two years, more rigorous requirements may be needed where exposure is in conditions that can be considered extreme, for example southern states of the USA.
- Where the pipe is stored outside for extended periods, the extent of degradation in storage could be assessed with an appropriate coloration and OIT testing, as discussed above.
- Mechanical performance in the form of tensile properties, impact resistance, fracture toughness, and burst stress, could be assessed as a function of exposure to elevated service temperatures and weathering. The duration of exposure should be such that the PE pipe is exposed to an equivalent dose of UV irradiation or elevated temperature. Artificial weathering and thermal exposures should only be performed when correlation to natural exposures has been proven, via physical assessments.

EFFECT ON WATER

Water quality issues are becoming a major concern to most water utilities. However, generally PE pipes have not caused major issue with water quality, except when buried in contaminated soils (Gaunt and Ong, 2005). It is probably the lack of issues that has led to some complacency in this area, although excellent standards such as ANSI/NSF 61 are available to be referenced. The lack of performance requirements at international level is understandable where ISO 4427-1 states that PE piping materials shall not constitute a toxic hazard, shall not support microbial growth and shall not give rise to unpleasant odour, cloudiness or discoloration of the water. ISO 4427-1 specifically draws attention to national regulations to address these issues. At the national level evaluation of performance can be assessed using individual national standards, such as BS 6920, AS/NZS 4020 or ANSI/NSF 61, that require testing according to methods specified by the regulatory body, and to levels of contaminants specified by the regulator or by the WHO guidelines, whichever is the strictest. Alternatively, evaluation can be performed using ISO 8795, which is also a CEN standard.

ANSI/NSF 61 is currently a voluntary consensus standard available for public use. When referenced by federal, state, or local health authorities in their specifications, piping manufacturers must have their products evaluated against ANSI/NSF 61, and the registered trademark of the testing agency indicating certification should appear on the PE piping products intended for use with potable water. ANSI/NSF 61 is very thorough in its requirement for health hazard evaluation, however as discussed by Burn et al. (2005), the extraction methodology could be simplified for time effectiveness without detriment to the hazard determination.

Ideally, assessment to a standard such as ANSI/NSF 61 should be carried out at both the compound and manufacturing level to ensure that firstly the compound does not affect the quality of water and secondly the manufacturing process also has no effect. In this respect standards such as the Australian standards require conformance to AS/NZS 4020 at both the compound (AS/NZS 4130) and manufactured pipe level (AS/NZS 4131).

In comparison, standards in the USA on the effect of materials on water quality are not rigorous. At the material level USA standards do not address this issue at all with there being no mention of water quality effects in ASTM 3350. At the pipe manufacture level the issue is only addressed in ASTM standards, with AWWA standards leaving the decision up to the purchaser as to whether the pipe needs to comply with ANSI/NSF 61. ASTM standards are more direct with a supplementary requirement being included that requires products intended for contact with potable water to be evaluated, tested and certified to conform with ANSI/NSF 61 or the health effects portion of NSF Standard 14. AWWA standards should at the minimum adopt this requirement for manufactured pipes.

INSTALLATION REQUIREMENTS

In many instances, PE pipe offers significant trenchless construction advantages over other pipe materials. While the later generations of PE materials are very tolerant to external damage, poor installation practices have contributed to some premature field failures in older PE materials with relatively low slow crack growth resistance.

Detailed methodologies for installation are generally provided by manufacturers and in many cases there are national standards in the area, For example the Australian standard AS 2033 for the installation of polyethylene pipe systems attempts to address some of these concerns.

Table 4.7
Trench conditions and dimensions

	AS 2033
Bed width	≥ 200mm greater than OD of pipe (pipe ≥ than DN125)
	≥ twice the pipe diameter (pipe smaller than DN125)
Underlay depth	Minimum 75mm
Pipe overlay	Minimum 150mm
Minimum cover over pipe	Subject to vehicle loading, 300mm
	Vehicle loading not in roadways, 450mm
	Vehicle loading in sealed roadways, 600mm
	Vehicle loading under unsealed roadways, 750mm
	Pipe in embankment conditions or subject to construction equipment loading, 750mm

Guidance and provisions referring to the transport, handling, storage and temperature limitations of polyethylene pipe are outlined in standards such as this. The area of most importance here is the pipe installation into the trench since most external pipe damage can occur at this point. The Australian standard requires that the following bedding material be used during trenching unless otherwise specified:

1. Suitable sand, free from rock or other hard or sharp objects that would be retained on a 13.2mm sieve.
2. Crushed rock or gravel of an approved grading up to a maximum size of 14mm.
3. The excavated material may provide a suitable pipe underlay if it is free from rock and free from any manner which has a diameter greater than 75mm.
4. The pipe side support and pipe overlay material shall be identical to the pipe underlay material.

In accordance with the Australian standard, Table 4.7 summaries the trench dimensions and conditions required to meet the specific requirements set out for polyethylene pipe installation practices.

In contrast to the Australian Standard, ASTM standard F645–03 outlines some brief installation guidelines with reference to buried pipe cover being not less than 300mm below the surface with a maximum cover of 600mm when pipe is subjected to heavy traffic loads. Both of these two particular requirements appear to coincide with that detailed in the Australian standard in Table 4.7. An important note in ASTM F645, is that if these requirements cannot be meet, precautions such as casing and bridging may also need to be used in conjunction with the required coverage depths.

Another ASTM standard D2774–04 also provides general recommendations with regard to underground installation of thermoplastic pressure piping. Even though the standard provides general guidance on trenching such as trench width, trench bottom, trench depth and pipe cover no specific requirements as indicated in Table 4.7 from the Australian standard are presented. As a result it is largely left to the installer or qualified engineer to specify the correct trenching conditions. However a minimum cover of 610mm is stated to be desirable with pipe subjected to overhead traffic, whilst a cover of 305 to 457mm is quoted for pipe subjected to light overhead traffic, both of these specifications full into the values outlines in the Australian standard.

A more informative standard ASTM F1688, construction procedures for buried plastic pipe is provided for open cut trenching taking into account the techniques and considerations of such an installation. Like all of the ASTM standards reviewed here the standard generally incorporates other pipe material such as PVC and ABS and is not in particularly focused on polyethylene pipe lines as a result.

At present the AWWA standard C901 contains a section which currently addresses the criteria for design and polyethylene pipe installation and in the future it is proposed that a separate standard solely addressing both pipe design and installation will be developed. Furthermore, the AWWA has recently issued Manual M55 on "PE Pipe—Design and Installation."

The guidelines and recommendations at present are very general with regard to trench conditions, installation practices and handling. The AWWA standard C906–2000 polyethylene pressure pipe and fittings for water distribution and transmission does not outline any practices or recommendations for pipe installation as such.

From the standards reviewed and studied in this section, it can be seen that the Australian standard outlines and describes the main requirements for an appropriate soil support system required when installing polyethylene pipe. Recommendations regarding the required pipe side support and pipe overlay and under lay are given in more detail than the ASTM and AWWA standards reviewed here. Both the AWWA and ASTM standards provide general recommendations with relation to the trench conditions but it is often left to the installer or engineer to provide the important recommendations and guidance in regard to soil depths and pipe side support, which are known to be critical in eliminating any possible pipe deformation from soil loading to a minimum.

Another important element in PE piping system installation is the proper joining of pipe to pipe and to fittings. The following standards and reports describe recommended procedures in detail:

- ASTM F1290, Standard Practice for Electrofusion Joining of Polyolefin Pipe and Fittings
- ASTM D2657, Standard Practice for Heat Fusion Joining of Polyolefin Pipe and Fittings
- TR 33, PPI Generic Butt Fusion Joining Procedure of Polyethylene Pipe (http://plasticpipe.org/pdf/pubs/reports/TR-33-05.pdf)

In contrast to countries such as Australia, in the USA and Canada, national training programs for installation and jointing of PE piping do not exist. Standard practices, such as ASTM F1290 ("Standard Practice for Electrofusion Joining Polyolefin Pipe and Fittings"), and guides offered by electrofusion fitting manufacturers give specific guidance for avoiding problems such as the ones that were featured in chapters 2 and 3.

National training programs do exist in other countries such as Australia, where short-term technical courses have been established for PE installations (http://www.ntis.gov.au; http://www.pipa.com.au/Training.html). These training programs are run by educational institutions and some water utilities require installation contractors to have undergone training before they will be awarded a contract. Proper training and qualification is required in both installation and jointing to ensure proper installation occurs and high quality heat fusion and electrofusion joints are created. Whilst M55 does highlight the importance of training and whilst manufacturers of jointing equipment do offer training programs, there is currently no nationally accredited scheme that utilities can require accreditation to and it is recommended that suitable training courses be established in the United States and that water utilities only use staff that are adequately trained in pipe installation and pipe jointing techniques. Another important factor is quality control and supervision and it is also recommended that water utilities should establish a quality control system that ensures that PE piping is properly installed in accordance with guidelines such as those specified in manual M55 "PE Pipe—Design and Installation."

CONCLUSIONS AND RECOMMENDATIONS

This chapter has reviewed and compared US and international performance standards for PE pipe systems. A comparison of material specifications indicates that AWWA and ASTM standards specify material property requirements for PE compounds such as density, melt index, flexural modulus, tensile yield strength and slow crack growth resistance. In contrast, ISO and Australian standards specify performance requirements for weathering resistance and slow crack growth resistance for a material manufactured in pipe form. Material properties such as density, yield strength and flexural modulus are not specified. However, there are some test criteria that should be addressed at the compound level that are not currently included in the US standards. For example, carbon black sizing, weathering resistance etc. are issues that may influence pipe performance, but are essentially compound characteristics and should be addressed at that level.

Both US and international standards provide similar methods to assess the long term hydrostatic strength of PE pipes. Each method differs in terms of data requirements, minimum test period durations and strength classification thresholds, but the main difference is in how the transition from ductile rupture failure to slow crack growth is handled. This transition in failure mode is reflected by a down-turn in the gradient of best fit lines to hydrostatic stress vs. time data. The ISO method has been developed to account for data in either ductile or slow crack growth failure regimes. As such, they require a relatively large number of data points to extrapolate to a long-term strength classification. In contrast, the ASTM method excludes those data points that fall in the slow crack growth regime. A separate requirement is then used to check for slow crack growth failures in long term data. It should be noted however, that for modern PE materials, with increased resistance to SCG, ambient temperature stress-rupture behavior follows a straight line demonstrating ductile rupture through at least the 50-year intercept. For these materials, both ASTM and ISO methods forecast essentially the same long-term strength.

Current short-term testing is variable and prescriptive, does not confirm long-term pressure performance and does not take material variation into account. It is recommended that short-term pressure testing to confirm the long-term HDB be adopted. This testing should allow for changes in material selection, and a suggested test procedure uses the 99.5% lower confidence limit value at 1 hour determined from the 50-year linear regression line for the appropriate material, determined using ISO 9080.

In designing for pressure transients (or surge) events in PE pipes, AWWA standards give the impression that a surge allowance has been incorporated into the pressure rating of the pipe. This impression is reinforced by the design factor of 0.5 used in these standards, compared to an ISO factor of 0.8, which effectively reduces the pressure class with the corresponding wall thickness penalties and increased costs incurred by the water utilities. Recent changes to the design factor to raise it to 0.63 for some resins, partially addresses this issue, however, these standards are still conservative, especially for the use of PE100 materials. It should also be noted that other researchers have reported that polymer pipes can become stiffer under higher rate pressurization (Marshall and Brogden 1997). As discussed by Burn et al. (2005) the practice of allowing for surge is questionable and for most pipelines operating under static pressure it is not needed. While US standards address the effect of recurring surge pressures, this methodology does not allow for de_hrating of PE pipes for different surge frequencies. Although they are not formally included in performance standards, UK and Australian industry guidelines for fatigue de-rating are available that account for surge amplitude, frequency and the required pipe lifetime. (PIPA 2002a) and an approach such as this may be worthwhile considering for incorporation into United States standards.

Whilst rapid crack propagation is not considered to be a major issue for water pipelines, ISO and Australian standards have addressed this issue for thicker pipes containing entrapped air. It is recommended that this potential failure methodology should be addressed by US standards developers, even to simply state that the potential for failure is considered low in warmer climates.

To assess the slow crack growth resistance of PE pipe systems, ASTM standards use a notched coupon specimen that is subjected to constant load at elevated temperature. In comparison, ISO and Australian standards require slow crack growth resistance to be demonstrated on notched pipe specimens. As outlined by Beech, Ferguson, and Clutton (2001), notched tensile coupon specimens provide a more severe test than the likely service conditions for a pipe. In particular, they do not account for the crack tip blunting mechanisms which are likely to occur in circumstances of low notch tip constraint (i.e., surface notches in pipes). Consequently, it is expected the relative benefits of the two test types will continue to be debated for some time and at the moment one test does not offer significant advantages over the other.

In the area of dispersion of additives, standards in the USA do not have a test requirement in this area and both ASTM and AWWA standards for polyethylene piping should consider adopting a methodology such as that specified by ISO to ensure adequate dispersion of additives in the manufactured pipes.

The stability of PE compounds under thermal and UV exposure depends on the stabilization system used, and the level of stabilizers incorporated. The level of stabilizers added is generally controlled by the thermal performance needs of extrusion, and in United States standards except for the inclusion of carbon black (for black pipes) little consideration is given to the long-term thermal performance requirements or UV performance requirements for manufactured pipes, apart from the generic requirement in AWWA C906 for pipes to contain sufficient stabilizers for 24 months of outdoor storage. ISO, AWWA and Australian standards such as ISO 4427 specify a minimum Oxygen Induction Time (OIT) for both compound and manufactured pipes to ensure that adequate heat stabilizers are incorporated into the resin and that these stabilizers have not been fully consumed during pipe extrusion. In comparison, ASTM standards only place a heat stabilizer performance requirement on the base compound of an induction temperature ≥220°C and have no requirement for manufactured pipe specimens. It is proposed that the different test methods adopted by USA and ISO standards should be compared to ensure that the performance requirements are aligned with each other. Additionally USA standards should consider adoption of a performance requirement for UV stability for pipes manufactured in colors other than black.

International standards for water quality issues in PE materials state that they shall not constitute a toxic hazard, shall not support microbial growth and shall not give rise to unpleasant odor, cloudiness or discoloration of the water. Ideally assessment for effect on water quality should be carried out at both the compound and manufacturing level to ensure that firstly the compound does not affect the quality of water and secondly the manufacturing process also has no effect. In ASTM standards water quality is addressed only at the pipe manufacture level and unfortunately with AWWA standards the purchaser decides whether the pipe needs to comply with a health hazard evaluation standard, such as ANSI/NSF 61. It is recommended that compliance with a standard such as ANSI/NSF 61 be compulsory at both the compound and manufactured pipe level.

As described in chapter 3, important issues influencing the service life of PE pipe systems are installation and jointing practices. In the US, standard practices for fusion jointing are available and it is recommended that water utilities establish a quality control system to ensure that PE piping is installed in accordance with these guidelines. National training programs exist in other countries such as Australia, where short-term technical courses have been established for PE installations. In some cases, water utilities require installation contractors to have undergone training before they will be awarded a contract. It is recommended that suitable training courses be established in the United States and that water utilities only use contractors that meet prescribed qualifications in pipe installation and pipe jointing techniques.

CHAPTER 5
SLOW CRACK GROWTH TESTING OF POLYETHYLENE

INTRODUCTION

Failure by the slow crack growth (SCG) process is the end result of gradual irreversible damage. The lifetime of a pipe depends on the rate at which this damage occurs. As the resultant cracks slowly grow, there is a proportional steady decrease in residual life. For those materials that are susceptible to SCG, it is essential to limit locally intensified stresses to a magnitude at which, under the conditions of service, the crack growth will proceed at such a low rate, or not at all, to result in sufficiently long service life. The brittle-like failures that occur in the ambient and elevated temperature stress-rupture testing of PE pipe are in response to localized stress-intensifications that are created by flaws that are inherent to the pipe material. However, projections of brittle-like long-term strength that are solely based on results of the stress-rupture testing of carefully prepared pipe do not adequately consider the possible adverse effect of "external" flaws (such as damage to the pipe outer surface). According to industry expert opinion, field experience indicates that external causes account more often that not for premature failure.

Previous research into the use of PE pipe for gas transportation has provided a range of different test methods to assess slow crack growth resistance, which also apply to water pipe materials. To rank the performance of different PE material grades, test methods that reproduce the SCG failure mode observed in the field within a practical time frame are essential. However, while a number of different test methods are available, the introduction of new generation PE materials (with improved SCG resistance) has resulted in impractically long test times in some cases. This chapter describes the methods of accelerated testing for slow crack growth resistance in PE materials.

ELEVATED TEMPERATURE HYDROSTATIC PRESSURE TESTING

As discussed previously in chapter 4, the long term performance of PE materials is classified in accordance with a Minimum Required Strength (MRS) category, which is determined from the results of hydrostatic pressure testing. For many PE materials, a transition is observed in hydrostatic pressure testing at elevated temperatures, as shown in the schematic diagram in Figure 5.1.

The variation of failure time with applied hydrostatic stress is divided into the regions, each corresponding to a different stage in failure mode. For stage I of hydrostatic pressure test data, the pipe is mainly subjected to mechanical overload and ductile failure occurs failure occurs by plastic instability. Ductile yielding is the end result of the gradual creep expansion of pipe which results in a thinning wall and a growing diameter, both of which contribute to the development of a steadily larger true hoop stress. The driving force is the bulk or, average value of the developed hoop stress.

In the transition to stage II, the time duration of the creep expansion and ductile failure processes observed in stage I is extended (due to the lower applied stress) and failures begin to occur by a quasi-brittle mode of slow crack growth. (Marshall, Pearson, and MacKellar, 1996). Stage II is often referred to as the "mechanical knee" in hoop stress vs. time data and corresponds

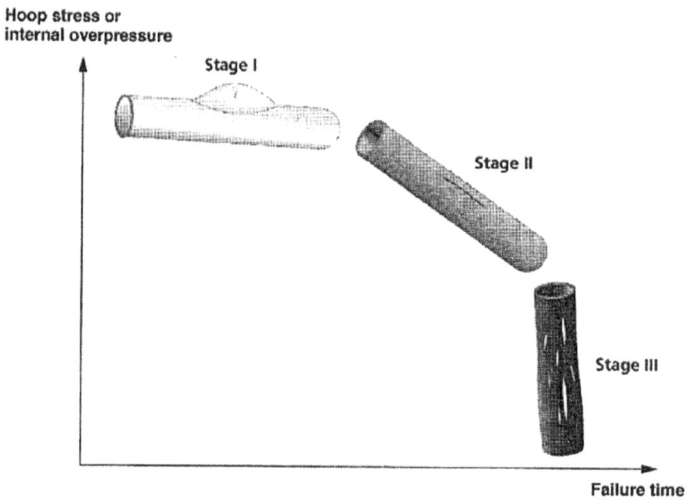

Figure 5.1 Schematic illustration of three stages during the service lifetime of a pressurized PE pipe (adapted from Andersson, 2001)

to more brittle-like failures which occur from inherent defects in the pipe wall at lower applied stress levels (Andersson, 2001).

In the later stages of this mechanical knee, a second transition is thought to exist to stage III, when the heat stabilizer package of the raw material is consumed and brittle fracture failures are induced by thermal degradation. Stage III is referred to as the "chemical knee" and failure times are effectively independent of applied stress levels (Andersson, 2001). Although Figure 5.1 represents the transitions in failure mechanism that can occur at elevated temperatures, it should be noted that it does not relate directly to possible field failures. For example, stage I is seldom, if ever, observed in field failures. While stage III is also not seen, there is one known exception. Pipe made by a single manufacturer, circa 1974, was found about 10 years later to become brittle, which caused the development of premature failures in water service applications. Laboratory tests indicated that the embrittlement was the result of oxidation due to a deficiency in stabilizer levels. This observation led to the establishment of the requirement, which has been included in AWWA and in other standards, that material taken from extruded pipe must meet certain minimum thermal stability criteria. Since then, there has been no report of a field failure that resulted by the Stage III process. Stage II is the dominant failure mode that has been observed to occur in the field.

Due to the relatively high slow crack growth resistance of pipe-grade PE materials, the ductile failure mode (stage I) is usually observed in hydrostatic pressure tests at ambient temperature. However, testing at elevated temperatures can be used to accelerate the failure transition process and observe stage II, slow crack growth failures within a practical time frame. As described previously in chapter 4, this also forms the basis of extrapolating short term data to give a long-term hydrostatic strength (and hence MRS categories). While hydrostatic testing is primarily used to assess long term hydrostatic strength of a PE material, the onset of the mechanical knee at elevated temperature qualitatively reflects the slow crack growth resistance of the material. Those materials which show the early onset of the knee (observed as a down turn in the stress vs. time data) at elevated temperatures would be expected to exhibit a relatively low slow crack growth resistance.

While accelerated testing at elevated temperatures is undoubtedly very useful, it is unclear if the actual crack growth mechanism at ambient temperature is preserved. For example, Barker, Bowman, and Bevis (1983) observed changes in peak melting temperature and percentage crystallinity in PE pipe samples after exposing pipes to a temperature of 80°C. Initially, pipe samples were held at 80°C for a range of test times with no load applied. As test time increased, gradual increases in peak melting temperature and crystallinity were observed, indicating that annealing had occurred. In a second assessment, an equivalent set of pipe samples were subjected to a fatigue loading cycle whilst being held at 80°C for the same test times. In this case, the addition of fatigue loading increased the rate at which melting point and crystallinity increased (Barker, Bowman, and Bevis, 1983). This is clear evidence that the microstructure of a PE pipe sample tested at 80°C differs from that of a pipe tested at ambient temperature and suggests that the mechanism of crack growth under ambient temperature is not preserved in tests at elevated temperatures. However, if an accelerated test procedure results in a ranking of materials that is equivalent to that achieved under ambient temperature (with significantly longer test times), then it is considered to exhibit great engineering value (Mruk, 2006).

NOTCHED SPECIMEN TESTING

To further accelerate crack growth in PE materials, testing of notched specimens under tensile stress is also used. An important reason for the use of notched specimens is to evaluate a material's resistance to SCG under the effect of "external" flaws (i.e., flaws other than those inherent to the pipe material).

For example, the Notched Pipe Test (NPT) was developed to provide a means of defining the brittle fracture performance of pipes in a reduced time scale. The test involves the hydrostatic pressurization at 80°C of a pipe which contains axial notches on the external surface at four points around the circumference. The NPT forms an international standard ISO 13479:1997: *Polyolefin pipes for the conveyance of fluids—Determination of resistance to crack propagation—Test method for slow crack growth on notched pipes* (ISO, 1997).

An alternative to pipe testing is the Full Notch Creep Test (FNCT), as outlined in the international standard ISO 16770:2004: *Plastics—Determination of environmental stress cracking (ESC) of polyethylene—Full Notch Creep Test (FNCT)* (ISO, 2004). In this test, a specimen in the form of a square section bar is either machined from compression molded sheet or from the pipe wall. Coplanar notches are introduced into each face of the specimen and the remaining ligament is subjected to a pre-defined static tensile stress. The test is performed in a detergent solution (such as Igepal) at 80°C to further accelerate crack growth. As outlined by Plummer, Goldberg, and Ghanem (2001), the presence of Igepal promotes the fine fibrillated structure on fracture surfaces, which is indicative of brittle crack growth.

Although both NPT and FNCT specimens use notched ligaments under applied static tensile stress, Beech, Ferguson, and Clutton (2001) note that different failures times are observed for the same applied stress. The data in Figure 5.2 compares stress vs. time to failure for a pipe grade PE material in the hydrostatic pipe test with data from the NPT and FNCT.

The solid lines in Figure 5.2 indicate data form hydrostatic pressure testing at 20, 60 and 80°C. While a ductile failure mode was observed in all tests at 20°C, a transition to brittle fracture was observed at elevated temperatures. This failure mode transition is reflected by the "knee" in the 60 and 80°C data. The solid circles in Figure 5.2 represent NPT data at 80°C and clearly show how failure is accelerated through the introduction of a notch in the pipe wall. According to

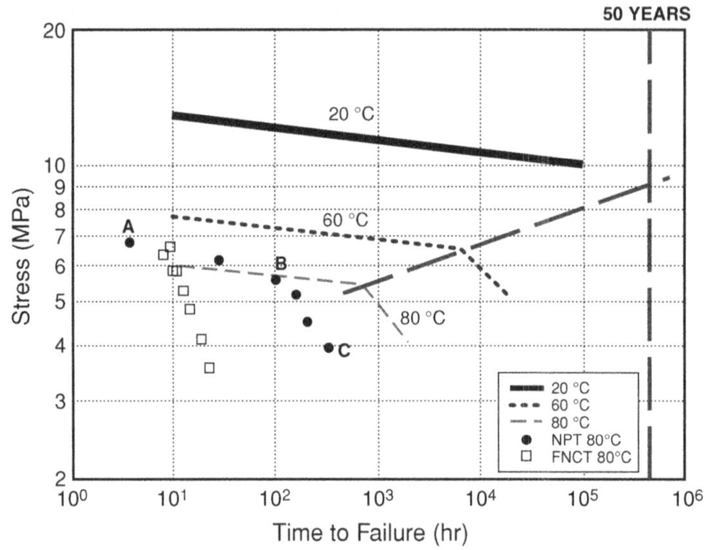

Figure 5.2 Comparison of stress vs. failure time data for hydrostatic pressure testing with NPT and FNCT on a pipe-grade PE material (adapted from Beech, Ferguson, and Clutton, 2001).

Beech, Ferguson, and Clutton (2001) a transition in failure mode is observed ranging from ductile crack tip blunting at high stress levels (denoted by point A in Figure 5.2) to the formation of multiple crack tip crazes at intermediate stress levels (point B) to the formation of a single crack tip craze at low stress levels (point C). In contrast, the FNCT data indicates further reductions in time to failure for a given applied stress. Microscopic examination of failed specimens indicate that crack tip blunting or multiple crazing was not observed in this test and all failures occurred by the development of a single crack tip craze. As concluded by Beech, Ferguson, and Clutton (2001), the FNCT provides a more severe test than the likely service conditions for a pipe. In particular, it does not account for the crack tip blunting mechanisms which are likely to occur in circumstances of low notch tip constraint (i.e., surface notches in pipes). As a consequence, the FNCT in isolation cannot be used to predict the likely performance in the NPT or the hydrostatic pressure test. However, it does define the inherent SCG resistance, which is part of the overall failure process. Beech, Ferguson, and Clutton (2001) also conclude that the crack tip blunting mechanism, which increases resistance to SCG, is promoted in newer higher density PE's that exhibit very high SCG resistance.

The Notched Tensile test, as outlined in the American standard ASTM F 1473 (ASTM International, 2001b), is similar to the FNCT. This test uses a specimen from a pipe with a specified orientation relative to the extrusion direction, or from compression molded plaque. It is precisely notched to a pre-defined depth and then loaded in tension to a pre-defined nominal tensile stress and the time to failure is recorded. Although no single test temperature is specified in ASTM F 1473, the test method is proposed to be useful for assessing slow crack growth resistance of PE materials under accelerated conditions (at elevated temperatures for example). The American standard ASTM D 3350 (ASTM International, 2002) specifies a test temperature of 80°C and a stress of 2.4 MPa. The ASTM notch tensile specimen differs from the FNCT in that a single deep notch is introduced on to one surface and shallower side notches on the two perpendicular faces. The face opposing the main notch remains un-notched. Previous research has shown that

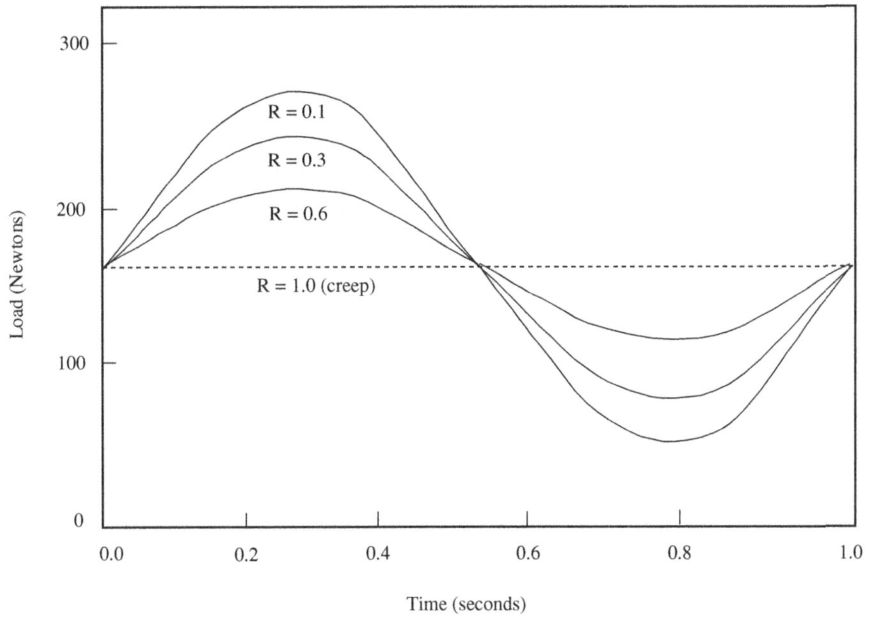

Figure 5.3 Fatigue loading for different *R*-ratios (adapted from Parsons et al., 2000)

the stepwise crack growth mechanism discussed earlier is also observed using this test method at elevated temperatures (Zhou and Brown, 1993). However, the increased crack growth resistances of newer PE materials mean that extended test times are required for failure, even at elevated temperatures (Zhou and Brown, 1993).

It is important to note that these empirical accelerated test methods can be a powerful tool in ranking the relative expected field performance of different PE materials. For example, Brown and Lu (1993) conducted SCG resistance tests on a range of PE gas pipe materials using the Notched tensile test in accordance with ASTM F 1473 (ASTM, 2001b). While no consistent relationship was observed, those materials that exhibited shorter failure times in the notched tensile test also generally exhibited higher average SCG failure rates in the field. Conversely, those materials with longer failure times generally exhibited higher average SCG failure rates in the field (Brown and Lu, 1993). It was also observed that for those materials exhibiting a lifetime greater than 14 hours, there were no reports of SCG field failures after 13 years of service. The current requirement for PE materials intended for gas distribution is a failure time of greater than 100 hours and it should also be noted that the newer generation PE materials exhibit PENT times that are in excess of 1,000 hours (Mruk, 2006).

CYCLIC LOADING TO ACCELERATE SLOW CRACK GROWTH

An alternative acceleration method that avoids the possible microstructural changes associated with elevated temperature is the use of fatigue testing. As described by Parsons et al. (2000), the relationship between fatigue and static loading response can be investigated by systematically decreasing the dynamic component of fatigue loading. In a series of experiments on a new generation medium density PE material, Parsons et al. (2000) used notched specimens machined from pipe to assess fatigue crack growth resistance under a range of loading regimes. By varying the minimum and maximum applied load, the dynamic component of fatigue was controlled as shown in Figure 5.3.

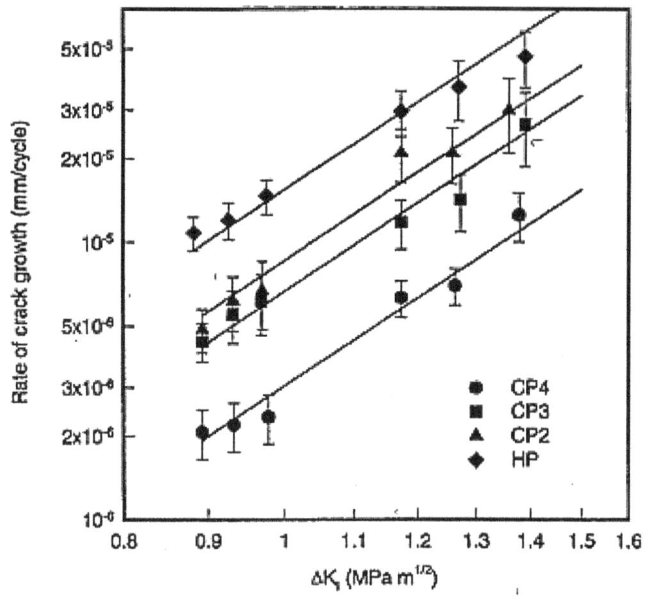

Figure 5.4 Fatigue crack growth data for different PE grades. *R*-ratio = 0.1 (data from Shah et al., 1998).

The *R*-ratio (ratio between minimum and maximum applied stress intensity factor) was gradually increased from 0.1 (fatigue loading) to unity (which corresponds to static loading). In this way, the effects of fatigue and creep were isolated and quantified. It is encouraging to note that relationships between applied stress and crack growth rates have been obtained for new-generation PE materials (with side branch comonomer distributions discussed previously) tested at ambient temperature. Shah et al. (1998) characterized fatigue crack growth resistance in a new generation bimodal PE material, for which test times under static load where impractically long, even at 80°C.

The example data in Figure 5.4 was obtained in a study by Shah et al. (1998) of the fatigue crack growth resistance in pipe-grade PE materials (Shah et al. 1998). As shown, for a given applied stress intensity factor range ΔK, each material exhibits a different crack growth rate under cyclic loading with an *R*-ratio of 0.1. The material designated "HP" was a PE homopolymer with no foreign side chain content. In contrast, material CP2 was a unimodal PE copolymer containing butyl short chain branches (SCB) assigned to low molecular weight PE chains. As shown in Figure 5.4, the fatigue crack growth resistance of CP2 is superior to that of the HP material. While the molecular weight of material CP3 was similar to that of CP2, its butyl SCB content (again assigned to low molecular weight PE chains) was increased. Consequently, its resistance to fatigue crack was increased. CP4 was a new generation bimodal PE material a broader molecular weight distribution and a concentration of butyl SCB in the high molecular weight part of the distribution. As shown in the Figure 5.4, the fatigue crack growth resistance of this material was highest of all materials tested. This superior performance was attributed to the increased number of inter crystalline links formed by tie molecules between crystalline lamellae (Shah et al. 1998).

Parsons et al. (2000) reported that discontinuous stepwise crack propagation in MDPE materials was observed in experiments under fatigue loading at elevated temperatures. However, as temperature decreased and/or *R* ratio increased (towards static loading), a transition was

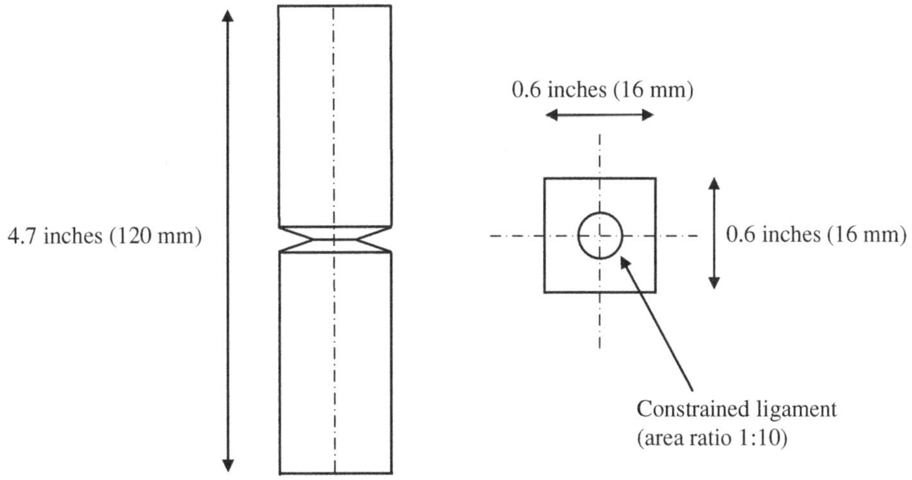

Figure 5.5 Schematic drawing of circumferentially deep notched tensile specimen for craze strength testing

observed from stepwise to quasi-continuous crack growth. This change was attributed to the reduction in strain rate within craze fibrils under static loading (Parsons et al. 2000) and suggests that although fatigue acceleration provides a quicker means of ranking the crack growth resistance of new bimodal PE materials, the mechanism of crack growth differs to that under static load conditions.

While fatigue testing at ambient temperatures can be useful for certain materials, tests show that test times are still impractically long. To achieve shorter test times, the Plastics Pipe Institute (PPI) has funded a program comparing results obtained using an elevated temperature fatigue test against those obtained by means of elevated temperature stress-rupture tests and also by means of the notched tensile test (as described in ASTM F 1473). When applying the two latter test methods on the newer, very high SCG resistant materials *accelerated* test times are found to be in the impractically long range of thousands of hours. In contrast, preliminary results indicate that the proposed elevated temperature *fatigue* test reduces the failure times by as much as two or three orders of magnitude. These preliminary results also show that the *same ranking of SCG resistance is established*. Therefore, although the precise mechanism of SCG may not be preserved at elevated temperatures, this should not rule out the use of such testing for an effective ranking of resistance to SCG.

CRAZE STRENGTH MEASUREMENT

In contrast to elevated temperature and fatigue acceleration, recent research has resulted in a new test method to isolate and measure the properties of the important crack tip craze, in pipe-grade PE materials (Pandya and Williams, 2000). The specimen geometry is shown schematically in Figure 5.5.

As shown, rectangular bar specimens are used with a symmetrical notch, machined around the circumference at the mid point and then razor-notched. The deep circumferential notch develops a highly constrained region in the small remaining ligament. As a consequence of this geometrical discontinuity, a tri-axial stress state and high magnitude of local stress is achieved (Pandya and Williams, 2000). Therefore, when loaded in tension, the high constraint promotes

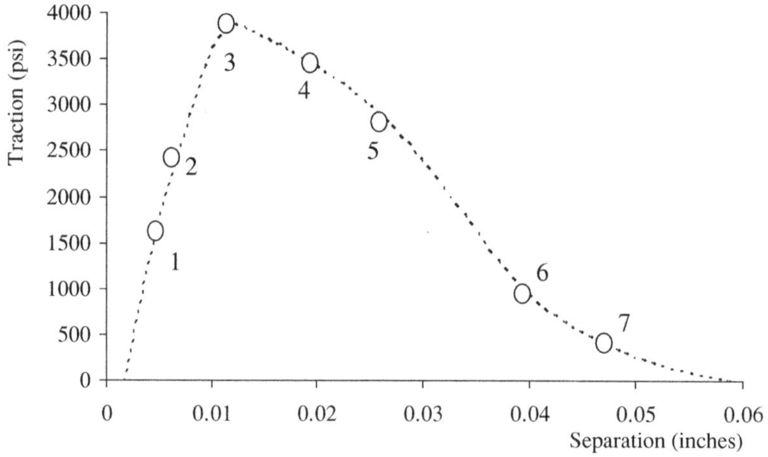

Figure 5.6 Traction-separation curve for a PE CDNT specimen at constant displacement rate of 0.0002 inches/min

craze growth in the ligament, which is the most severe damage mechanism likely to occur for a flawed pipe in service.

Circumferentially Deep Notched Tensile (CDNT) specimens are tested under constant displacement rate or constant load conditions. To measure the total craze extension, a clip gauge transducer is mounted across the separating surface. The specimen was pin mounted in universal joints at each end to prevent bending and twisting in the specimen. Load is measured using a standard load cell. Testing at constant displacement rate allows traction-separation curves to be generated, where traction is calculated as the recorded load divided by the ligament cross sectional area. An example curve for an Australian PE material is shown in Figure 5.6. The images in Figure 5.6 indicate the individual stages of craze formation, growth and eventual failure.

As shown in Figure 5.6, the traction-separation curve initially increases linearly. At this stage in the process, the material is homogeneous and the craze damage zone has not yet formed. At point 1, the traction-separation curve becomes non-linear corresponding to uniform craze formation across the ligament as shown in Figure 5.7, image 1. At point 3, the maximum traction is attained and voids are beginning to form in the crazed ligament (Figure 5.7, image 3). This corresponds to the craze strength σ_D defined as the maximum load divided by the ligament cross sectional area. After peak traction is attained, the size and number of voids increases as the craze extends with the traction decreasing. This corresponds to subsequent damage evolution in Figure 5.7 (images 4 to 7).

In addition to testing under constant displacement rate conditions, the time dependence of craze strength can be obtained by conducting tests under static load conditions. For example, the data in Figure 5.8 shows the variation in time to failure for a range of different CDNT specimens under static load.

Material PE4 was a homopolymer with short chain branch (SCB) content and similar to the crack growth study by Shah et al. (1998), this material exhibited the lowest craze strength for a given failure time. In contrast, material PE3 was a copolymer, with a relatively low SCB content, and exhibited increased craze strength. However, no information was provided describing the location of SCB (i.e., assigned to low or high molecular weight PE chains) on this material. Materials PE2 and PE1 were also copolymers but exhibited different molecular weights and SCB

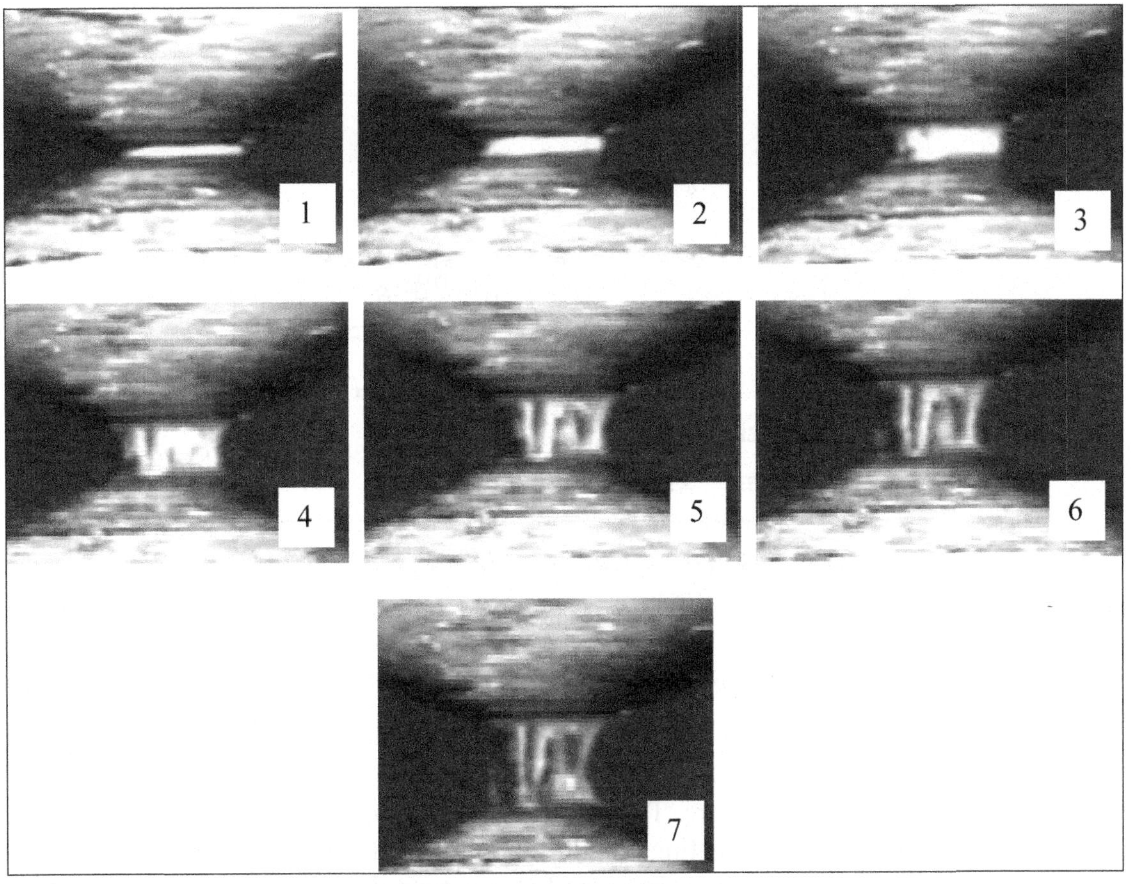

Figure 5.7 Stage-by-stage images of craze formation, growth and separation in a PE CDNT specimen. Constant displacement rate of 0.0002 inches/min.

contents. While material PE2 exhibited a moderate SCB content, its molecular weight was highest of all materials tested. In contrast, material PE1 exhibited a moderate molecular weight but the highest SCB of all materials tested. As shown in Figure 5.8, both these materials exhibit relatively high craze strengths compared to the other materials tested. This suggests that both high molecular weight and high SCB content increase long term craze strength in PE materials. However, as with materials PE4 and PE3, the location of SCB in PE2 and PE1 was not provided.

CONCLUDING REMARKS

Hydrostatic pressure testing of PE materials usually produces a ductile failure mode at ambient temperature. At elevated temperatures, a transition to a brittle-like slow crack growth (SCG) failure mode can be observed (depending on the material). The early onset of this transition indicates a material with relatively low SCG resistance. In addition to elevated temperature hydrostatic pressure tests, a range of test methods which use notched specimen geometries are also available for SCG resistance assessment. In general, the performance of older PE materials can be assessed within practical time frames and ranked consistently between the different test methods available. However, evidence in the literature suggests that material microstructure and the mechanisms of crack growth at elevated temperatures may depart from those experienced by a

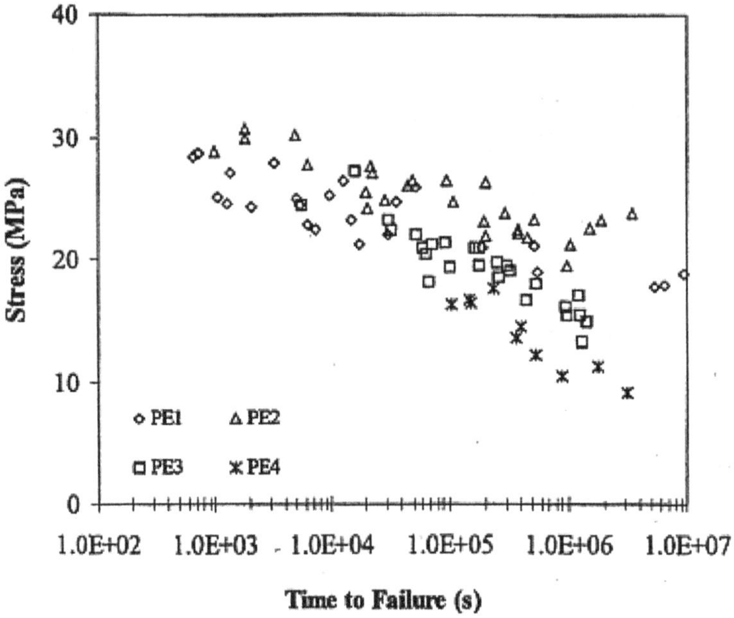

Figure 5.8 Craze stress vs. time to failure data for different PE materials (data from Pandya and Williams, 2000)

pipe in service at ambient temperature. Therefore, for the purpose of pipe service life prediction, results from elevated temperature testing should be interpreted carefully. For relative ranking of the SCG resistance of different PE materials within practical timeframes, these test methods are invaluable engineering tools.

While the performance of older PE materials can be assessed using elevated temperature and notched specimen tests, continuous improvements to SCG resistance mean that for new PE materials, test times are impractical even at elevated temperatures. Recent advances in SCG testing have led to the use of cyclic loading to accelerate the formation and failure of the crack tip craze zone in PE. While there is evidence to suggest that the mechanism of crack growth differs from that observed for pipes in service, qualitative ranking of new PE materials with high SCG resistance has been possible in practical time frames. A novel test method has also recently been developed which isolates the craze failure process in PE materials. However, while this method also provides a ranking for materials in terms of their inherent resistance to slow crack growth, the basis for translating results to service lifetime predictions for PE pipes is unclear.

CHAPTER 6
PREDICTING LONG-TERM PERFORMANCE IN POLYETHYLENE PIPES

INTRODUCTION

Although theories have evolved to give a fairly good account of the nature of the slow crack growth (SCG) process it is important that results predicted by a particular theory be calibrated against actual experience. To accomplish this, the factors that govern field failures must be identified, understood and periodically compared against actual field performance. This objective has been largely realized due to the high performance demands of the gas distribution industry. After the SCG mechanism was identified as the major determinant of lifetime in a PE pipe, concerted efforts were expended in developing measures for SCG resistance, for improving the SCG resistance of PE materials and for effecting changes in the standards that would exclude low SCG resistant materials from piping applications. There were also efforts for comparing the quality of field performance against a measure of SCG resistance which resulted in the establishment of minimum SCG resistance requirements. Therefore, it should be noted that while theoretical models (based on fracture mechanics for example) are useful, empirical methods based on experimentally measured SCG resistance also have proven merit in estimating service lifetime (Mruk, 2006). This chapter examines empirical and theoretical methods that can be used to assist predictions of long-term SCG failure of PE pipes.

RATE PROCESS METHODS

The Rate Process Method (RPM) is used for forecasting from shorter-term elevated temperature stress-rupture test results, the stress-rupture behavior that will occur under similar conditions at lower temperatures but after substantially longer times. The theoretical basis for the Rate Process Method (RPM) is the rate at which energy can be supplied to regions where energy barriers impede a rate process. Coleman (1956) noted that the magnitude of an activation barrier should be a function of stress, and that the rate at which a process proceeds should be described by

$$R = \frac{KT}{\eta}\exp\left[-\left(\frac{\varepsilon - \frac{\beta\sigma}{2}}{KT}\right)\right] \tag{6.1}$$

where R is the rate at which activated units (molecules) 'jump' the energy barrier, T is the absolute temperature, σ is the true stress at the activation barrier (i.e., for fracture, σ is the true stress at the crack tip), ε is the activation energy and the remaining parameters are constants. Assuming a critical strain at fracture, Coleman (1956) derived an expression for failure time t from Equation 6.1.

$$\log t = \frac{A - \sigma}{BT} \tag{6.2}$$

where A and B are constants. The difficulty in using Equation 6.2 for pipes in practice is that the true stress at a crack tip cannot be defined. Consequently, empirical relationships have been established between t, T and σ, which follow the form of Equation 6.2 but replace true stress by the nominal stress. For example, in a pipe under internal pressure, σ would be the applied hoop stress in the pipe wall. The widely used form of these relationships is a log-linear combination of the effect of temperature and hoop stress on failure time (Palermo, 2004).

$$\log t = A + \frac{B}{T} + C\frac{\log\sigma}{T} \tag{6.3}$$

where t is failure time, T is temperature and σ is hoop stress. This is an Arrhenius representation in which the activation energy of the failure process varies linearly with the logarithm of applied stress (Popelar, 1993).

The empirical constants A, B and C must be determined from regression fits to experimental burst data on pipe under internal pressure. The fitting of Equation 6.3 requires failure time data from a minimum of two temperatures. The data from these temperatures must then follow a linear dependency when plotted in the log t vs. log σ plane. If a transition from ductile failure (at high stress) to brittle failure (at low stress) occurs, then this fitting procedure must be followed for both failure regimes (i.e., log t vs. log σ data must be generated for at least two temperatures in both failure mode regions. Extensive tests on PE pipes in the US at various temperatures and pressures have shown that the RPM Equation 6.3 provides a good fit to experimental data (Palermo and DeBlieu, 1985). Having obtained the empirical parameters A, B and C, further statistical analysis indicates that Equation 6.3 can be confidently extrapolated to determine conditions for longer-term pipe failure (Palermo and DeBlieu, 1985). An additional suggested advantage of RPM testing is that experimental data must be generated at higher temperatures up to 95°C. This ensures that the observed failure mode is one of brittle fracture, similar to that observed in the field. Short-term hydrostatic pressure tests conducted at lower temperatures produce a ductile burst failure mode, which is rarely observed in practice. As discussed by Palermo (2004), RPM test methods have also been developed to predict failure times for PE pipes under additional, secondary loads such as rock impingement, squeeze-off and bending. By testing exhumed sections of PE gas pipe, Palermo (2004) indicated that RPM projections of service lifetime correlate well with actual field performance (expressed as an average failure rate per pipe length per year) of PE gas pipe installed in the US. In general, those pipes that exhibited shorter projected failure times from RPM testing also exhibited higher failure rates in service.

However, it should be understood that while the RPM is based on the hypothesis of a continuous process this assumption does not take into account the incubation and the subsequent stop and go steps by which slow crack growth failures occur. However, the equations that have been developed based on the RPM process have been found to fit the experimentally obtained data quite well. So, while the RPM process is founded in a certain theory that is not exactly followed, its effectiveness is supported by experimentally obtained data (Mruk, 2006).

The Applicability of Rate Process Methods

Whilst rate process methods specify accelerated testing at elevated temperatures, it is unclear how the transition from ductility to brittle fracture failure in these tests relates to the conditions that a PE pipe experiences in service. As suggested by Marshall, Pearson, and MacKellar

(1996), accelerating brittle fracture from extremely small inherent defects in a pipe (introduced during manufacture) by increasing temperature is a situation that does not occur in practice. However, it should be recognized that the quality of resistance to failure that is caused by internal defects as assessed from elevated temperature tests) is also an index of a material's resistance to external defects. Materials that are predicted to exhibit a very late downturn in their ambient temperature stress-rupture behavior have been found to perform much better under field conditions, when compared to materials that exhibit an early downturn (Mruk, 2006).

Although no details are given, Marshall, Pearson, and MacKellar (1996) also argue that the mechanisms of cracking in PE pipes at 80°C differ to those occurring in pipes operating at lower field temperatures. As outlined in chapter 5, other researchers have also proposed that the material microstructure changes as temperature is increased (Barker, Bowman, and Bevis, 1983). However, as discussed previously, a change in microstructure at elevated test temperatures should not automatically negate the value of elevated temperature testing. As long as an empirical correspondence exists between the accelerated results at elevated temperatures and the longer time results at ambient temperatures, there is value in such testing to rank the performance of different PE materials. As suggested by Marshall, Pearson, and MacKellar (1996), an alternative to elevating temperature is to promote embrittlement by deliberately introducing larger cracks in test samples, which suppress the plastic phase of creep deformation in long-term pressure tests. This is thought to represent the more realistic situation of pipe damage incurred during handling and installation. However, when such tests are conducted at room temperature on the newer materials that exhibit very high resistance to SCG, testing times can still be un-practically long (Mruk, 2006).

BIDIRECTIONAL SHIFTING OF HYDROSTATIC STRENGTH DATA

Originally based on stress relaxation tests, Popelar, Popelar, and Kenner (1990) developed a method to transform results from high temperature testing of PE to data points at ambient temperature. The so-called bidirectional shifting method applies a horizontal transformation to time data and a vertical transformation to the measured quantity of interest. After examinations of around 45 datasets for US PE materials, a constant set of bidirectional shift factors were found to apply to the time/temperature dependence of crack initiation, crack growth rate and the failure time in hydrostatic strength tests on PE pipes, single edge notched tensile specimens and plane strain tensile specimens (Popelar, Popelar, and Kenner, 1990). For example, consider a hydrostatic strength test conducted on a PE pipe in the laboratory at some elevated test temperature T_t and applied stress σ, which results in an accelerated failure time t. Following Popelar (1993), this data point can be mapped to a corresponding point at the reference temperature T_s (corresponding to the actual temperature of the pipe in service for example). The mapped point at reference temperature T_s is at a stress level σb_T and a failure time t/a_T. The parameters a_T and b_T are bi-directional time-temperature shift factors given by

$$a_T = \exp[0.109(T_s - T_t)] \tag{6.4}$$

and

$$b_T = \exp[0.0116(T_s - T_t)] \tag{6.5}$$

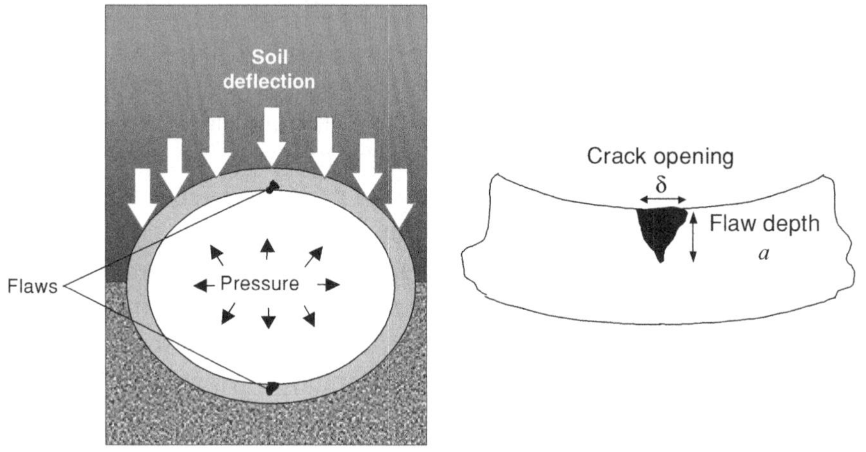

Figure 6.1 Expected operating conditions of a buried plastic pipe

where T_s and T_t are the actual service temperature of the pipe and temperature at which data was obtained in the lab. As stated above, Equations 6.4 and 6.5, which were originally based on stress-relaxation data, were also found to apply to apply to crack initiation and slow crack growth data as well as failure time data for US PE pipe materials (Popelar, Popelar, and Kenner, 1990).

The Applicability of Bidirectional Shifting

A comparison with the Rate Process Method suggests that for fewer experimental data points, the bidirectional shifting method may provide a more accurate representation of the time/temperature dependence of hydrostatic strength data for US PE pipes (Popelar, 1993). It is also noted that the bidirectional shifting factors in Equations 6.4 and 6.5 apply to both ductile and brittle pipe failure modes. In comparison with the increased experimental data requirements of the Rate Process Method to separately represent both ductile and brittle failure modes, bidirectional shifting is proposed to be a more cost efficient method (Popelar, 1993). However, although the bidirectional shifting method is proposed to be universal (applying to the full range of visco-elastic processes that occur), it is an empirical method. As reported by Popelar (1993), the approach has been viewed with some skepticism since the Arrhenius dependency of stress on time and temperature (as indicated by the Rate Process Method) is not observed.

LINEAR ELASTIC FRACTURE MECHANICS (LEFM)

In the case of brittle fracture, Linear Elastic Fracture Mechanics (LEFM) theory idealizes the fracture process in a plastic pipe as a balance between an applied stress intensity factor (SIF) and the fracture toughness of the pipe material (Williams, 1978). The applied SIF is determined by in-service loading conditions and the "current condition" of the pipe. Service loading can be split into contributions from operating pressure, external soil loading, residual stress, and external impact. The current condition of a pipe is characterized by damage (i.e., external surface scratches and inclusions) in the pipe wall. A possible set of conditions that could lead to fracture failure in service is illustrated schematically in Figure 6.1.

The inner surface of a buried plastic pipe is subjected to tensile stresses from internal pressure and bending stresses from soil deflection loads. Consider a flaw at the crown or invert of the pipe. The magnitude of applied stress will increase at the tip of a sufficiently severe flaw and crack initiation will eventually occur. This crack will slowly grow through the wall thickness until at some point, the applied SIF exceeds the fracture toughness of the pipe material and brittle fracture will occur. Although it is a less severe case, an external surface flaw at the springline of the pipe would grow radially inward eventually producing brittle fracture. However, it should also be noted that, as discussed in chapter 1, the residual stress at the outer surface of an extruded PE pipe is compressive in the circumferential direction. Consequently, crack initiation from a defect located at the outer surface of the pipe is less likely to occur.

In addition to those stresses illustrated in Figure 6.1, it should also be noted that a buried pipe is subjected to other localized stress increasing situations (i.e., rock impingement, gouges, differential settlement, poorly made fusions, sudden changes in geometry) which are major factors in the development and propagation of slowly growing cracks. A PE pipe must exhibit high resistance to the potentially adverse effect of all "normally" occurring localized stress increasing situations.

According to LEFM theory, the fracture process can be split into three stages, which combine to determine the total failure time under a static load (Williams, 1978)

1. An incubation period between load application and crack initiation.
2. A period of crack growth.
3. Fast brittle fracture when the applied SIF exceeds the material fracture toughness.

However, as outlined by Sandilands and Bowman (1986), slowly propagating cracks have been observed to occupy 90% of the wall thickness in both MDPE and HDPE pipes, with final rupture occurring by a ductile mechanism. This suggests that incubation and slow crack growth (stages 1 and 2 only) determine lifetimes in pipe-grade PE.

To calculate service lifetimes associated with stages 1 and 2, the stress intensity factor (SIF) associated with a flaw in the pipe wall must be quantified. If a flaw of radial depth "a" exists in the wall of a buried pipe a SIF, (K) can be defined, which relates the remotely applied stress σ in the pipe wall to the stress at the flaw tip.

$$K = \sigma Y(\pi a)^{1/2} \qquad (6.6)$$

Y is a geometric correction factor, which accounts for the influence of pipe geometry and loading type on K. Geometric correction factors have been previously defined for plastic pipes under combined internal pressure and diametrical deflection (Lu, Davis, and Burn, 2004). Solutions are also available in the literature for processed-in residual stress (Lu, Davis, and Burn, 2004). For a buried pipe, a total applied SIF can therefore be defined as

$$K_{ITOT} = K_{IP} + K_{ID} + K_{IR} \qquad (6.7)$$

where K_{IP} is the applied SIF from internal pressure, K_{ID} is the SIF from deflection and K_{IR} is the SIF for residual stress (Lu, Davis, and Burn, 2004). It should be noted that for simplicity, Equation 6.7 only includes the effect of those stresses depicted by Figure 6.1. As stated above, other conditions can be present in actual buried pipe situations that produce a possibly greater total stress intensity factor than that which is predicted. Previous researchers have proposed that the incubation period before crack initiation ends occurs when the defect or crack opening displacement (COD) attains a critical value (Williams 1978, Marshall 1982). If visco-elastic behavior is represented by the time-dependant Young's modulus $E = E_0 t^{-p}$ and $\sigma_y = \sigma_0 t^{-q}$, then the incubation time for crack initiation (t_i) can be written as:

$$t_i = \left[\frac{E_0 \sigma_0 \delta^*}{K_I^2}\right]^{\frac{1}{p+q}} \qquad (6.8)$$

However, previous research has shown that Equation 6.8 can significantly over-predict initiation times at low applied stress intensity factors (Burn et al. 2005). After incubation ends, a period of crack growth occurs. Under a static load, fracture mechanics theory proposes that K can be related empirically to crack growth rate da/dt in the form,

$$\frac{da}{dt} = DK^m \qquad (6.9)$$

where D and m are constants, which can be obtained from laboratory fracture tests (Sandilands and Bowman, 1986). Assuming that the crack grows in a stable manner through the pipe wall, Equation 6.8 can be integrated between the limits of defect size and pipe wall thickness to obtain the time to rupture. It is also assumed that the geometry factor Y in Equation 6.6 remains constant during crack growth. Sandilands and Bowman propose that under internal pressure only, this is reasonable since the majority of time to rupture is spent with the crack length small (less than ½ the pipe wall thickness), until the pipe wall is breached in the final stages of lifetime. Under these conditions, Y is effectively constant (Rooke and Cartwright, 1976) and integrating Equation 6.9 between the limits of initial defect size and pipe wall thickness gives the time to stress rupture τ_{SR} as

$$\tau_{SR} = \left[\frac{2(\pi^{1/2} Y \sigma_H)^{-m}}{D(2-m)}\right]\left[b^{1-(m/2)} - a_o^{1-(m/2)}\right] \qquad (6.10)$$

σ_H is the hoop stress in the pipe wall, b is the pipe wall thickness and a_o is the initial defect size. It can be shown that if b is much greater than a_o and m is sufficiently large, then Equation 6.10 reduces to

$$\tau_{SR} = \left[\frac{2(\pi^{1/2} Y \sigma_H)^{-m}}{D(m-2)}\right] a_o^{1-(m/2)} \qquad (6.11)$$

Therefore, the crack growth exponent m can be obtained experimentally by subjecting pipes with different inherent defect sizes to internal pressure and recording times to failure. For a given inherent defect size, m is obtained form the gradient of a $\log(\sigma_H)$ vs. $\log(\tau_{SR})$ plot. In examining the role of flaw size on crack growth rates, Sandilands and Bowman (1986) tested HDPE pipes which were extruded with deliberately added spherical aluminum particles of different diameters.

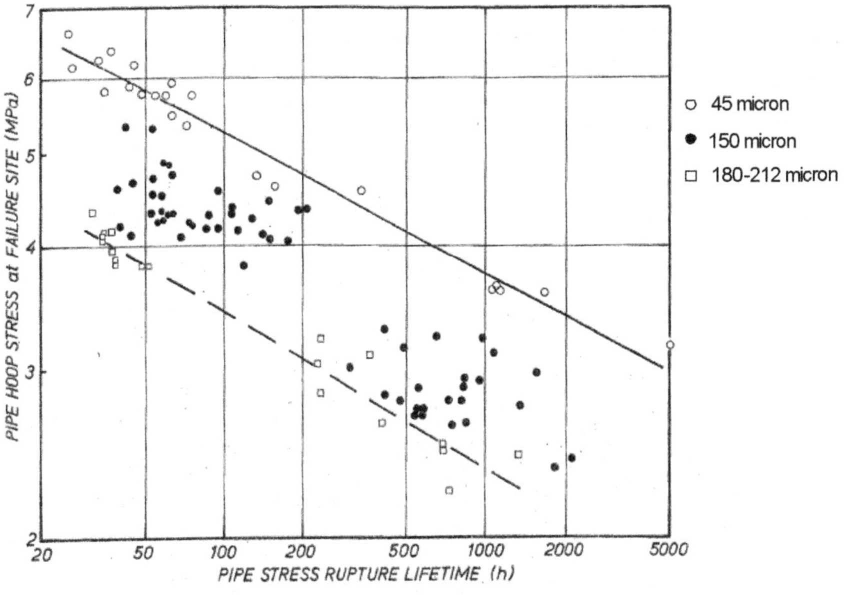

Figure 6.2 Expected stress rupture data for HDPE pipes with different defect sizes (data from Sandilands and Bowman, 1986)

Example, $\log(\sigma_H)$ vs. $\log(\tau_{SR})$ plot from the study by Sandilands and Bowman (1986) are illustrated in Figure 6.2.

As shown, although larger defect sizes reduce τ_{SR} for a given applied stress, the gradient of each plot is similar. Analysis of this data gave m-values between 6.25 and 6.67 for defects between 45 microns and 212 microns in diameter (Sandilands and Bowman, 1986). This suggests that while larger defects increase crack growth rate, the crack growth exponent m is independent of defect size.

Assuming that crack incubation periods are small, and that lifetime is governed by slow crack growth only, Equation 6.11 can also be re-cast in the form

$$\log(\tau_{SR})_1 - (1 - m/2)\log(a_o)_1 = \log(\tau_{SR})_2 - (1 - m/2)\log(a_o)_2 \tag{6.12}$$

where the subscripts 1 and 2 denote different defect sizes. Therefore, for a given pipe size and hoop stress, Equation 6.11 allows the lifetime for a defect of size $(a_0)_2$ to be predicted from measured lifetimes corresponding to defect size $(a_0)_1$. To validate this assumption of small incubation periods, Sandilands and Bowman (1986) used Equation 6.11 to successfully predict lifetimes for pipes containing defects of 200 micron in diameter, based on measured lifetimes in pipes containing defects 45 micron in diameter. However, while this result suggests that for these defect sizes, incubation periods are small, other researchers have reported extensive incubation periods in pipe-grade PE samples, which govern the time to failure (Bragaw, 1978).

To account for the effect of cyclic loading on crack growth rate, a generalized version of Equation 6.9 was proposed by Parsons et al. (2000) in the form

$$\frac{da}{dt} = D'K_{I,max}^{m}(1 + R)^{n} \tag{6.13}$$

Table 6.1
Slow crack growth parameters for PE materials. K_I expressed in MPa $m^{0.5}$ Crack growth rate *da/dt* expressed in mm/sec × 10^{-5} (adapted from Parsons et al., 2000)

Material	m	n
HDPE (USA)	4	–0.5
MDPE (USA)	4	–6

In Equation 6.13, K_{Imax} is the maximum applied SIF during cyclic loading and R is the ratio between maximum and minimum applied SIFs (for static loading conditions, $R = 1$). n is an empirical parameter that describes the fatigue acceleration in crack growth due to cyclic loading. Under static loading conditions, the right hand side of Equation 6.12 reduces to $(2^nD')K_I^m$, which is comparable to Equation 6.8. In a study of MDPE and HDPE materials in the US, Parsons et al. (2000) obtained values for crack growth parameters m and n as shown in Table 6.1.

An m-value of 4 as shown in Table 6.1 suggests that the slow crack growth resistance of US MDPE and HDPE material is higher than those tested by Sandilands and Bowman (1986), which exhibited m-values between 6.25 and 6.67. However, values of the D parameters in Equation 5.12 were not available for comparison. It is also interesting to note that the fatigue acceleration parameter, n, differs for each material in Table 6.1. Referring back to Equation 6.13, the value of $n = -6$ for MDPE suggests that this material has a significantly higher resistance to fatigue crack growth than HDPE. However, in the absence of molecular weight and comonomer distribution details, it is unclear whether this difference can be attributed to inherent differences in material microstructure.

LEFM-Based Service Life Prediction Tools for PE Gas Pipes in the US

As discussed in chapter 1, the majority of research into long term performance of PE pipes has been driving by the gas industry, where PE pipes are almost exclusively used. This research has led to lifetime prediction models that (providing the same materials are used) can also be adapted to service life prediction for PE water pipes. For example, LEFM theory is used in LIFESPAN (Life Integrity Forecasting and Enhanced Service Performance ANalysis), which is the service life prediction software for PE gas pipes developed by Southwest Research Institute and the Gas Research Institute in the US (Kuhlman et al. 1997). As described by Kanninen, Kuhlman, and Mahmoun (1993), LEFM analysis of extensive slow crack growth tests on PE gas pipe materials provides an equation for time to failure t_f given by

$$t_f = a_T\left[b_T^n B K_o^m + \frac{1}{b_T^m A}\int_{a_o}^{h}\frac{da}{K^m}\right] \tag{6.14}$$

a is instantaneous crack length, a_o is the initial crack length and h is the pipe wall thickness. K_o and K are the initial and instantaneous applied Stress Intensity Factors respectively. These K-values include contributions from internal pressure and residual stress in the pipe wall from outer surface cooling (Kanninen, Kuhlman, and Mahmoun 1993) The empirical parameters A, B, m and n describe the material resistance to crack initiation and slow crack growth and are obtained by slow crack growth (SCG) experiments on single-edge-notched bend specimens subjected to different static loads. Kanninen, Kuhlman, and Mahmoun (1993) document values of these SCG parameters for different PE material grades used for PE gas pipes in the US. a_T and b_T are bidirectional shift factors as defined previously in Equations 6.4 and 6.5.

The LIFESPAN program is menu driven and prompts the user for the slow crack growth material properties (as documented in Kanninen, Kuhlman, and Mahmoun (1993)) and the operating conditions of the pipe of interest (e.g., diameter, SDR, operating pressure, service temperature). LIFESPAN then evaluates Equation 6.13 and outputs the predicted time to failure. Parametric studies can also be conducted to determine the effect of variables such as initial crack length, residual stress, service temperature and operating pressure on predicted lifetime. LIFESPAN also offers a number of post-processing options such as data-viewing and plotting for further analysis (Kuhlman et al. 1997).

The Applicability of LEFM to PE Materials

While LEFM theory has been used to successfully predict failure times in some pipe-grade PE materials, its validity has been questioned for others. For example, using an MDPE pipe material from the US, Bragaw (1978) demonstrated that the incubation period before crack initiation actually increases for higher applied SIFs. Clearly this behavior contravenes the validity of LEFM theory for these materials. It was proposed that the reason for this increase was local visco-plastic deformation and blunting of the crack tip under high applied stresses. This would inhibit crack initiation and subsequent brittle fracture (Bragaw, 1978). As discussed in chapter 3, Plummer, Goldberg, and Ghanem (2001) also report that at high stresses, failure of notched tensile specimens occurs by ductile yielding.

The metric for deciding the applicability of LEFM theory is the extent of crack tip plastic deformation that occurs during fracture property testing. If the stress perpendicular to the crack is σ_{yy} and the distance from the crack is r, LEFM theory predicts that

$$\sigma_{yy} = \frac{K}{(2\pi r)^{1/2}} \tag{6.15}$$

where K is the applied stress intensity factor. As shown in the schematic diagram in Figure 6.3, a stress singularity is observed as r approaches zero, which is characteristic of LEFM theory.

According to the ($r^{-1/2}$) dependency in Equation 6.14, LEFM theory predicts that the stress right at the crack tip is infinite (Atkins and Mai, 1988). However, if this were true, bodies containing cracks would be unable to sustain any loads and have no strength at all. In reality, high crack tip stresses in PE materials are relieved by craze formation (over some length as shown in Figure 6.3) and the stress singularity is removed. However, LEFM theory ignores the details of events that take place in the craze zone and simply uses a scaling parameter (the Stress Intensity Factor K) of the stress field outside the craze zone to characterize the fracture behavior (Duan and Williams, 1998). This is valid when the craze zone is sufficiently small to be engulfed in a stress field dominated zone, and all microscopic events happening in the craze zone can be regarded as being controlled by the macroscopic stress field. However, difficulties occur for tougher PE materials which produce larger craze zones. As discussed in chapter 3, the separation of these craze zones absorbs a significant amount of work available for crack propagation and confers high

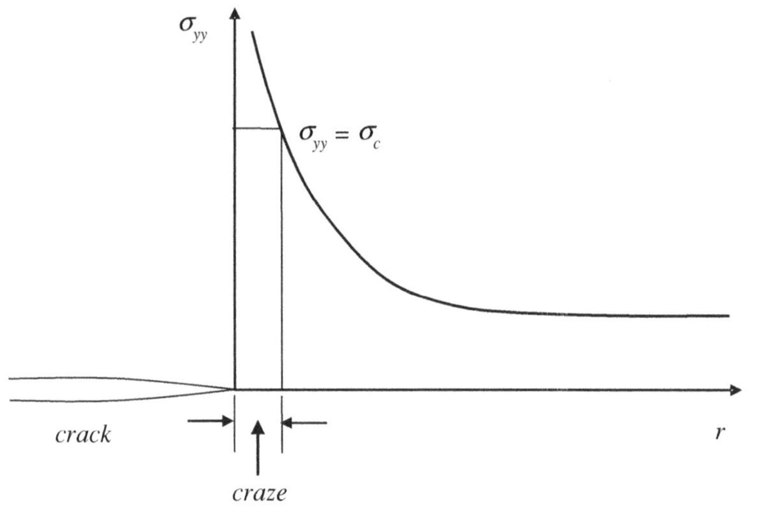

Figure 6.3 ***K*-dominant stress field and crack tip singularity according to LEFM theory**

toughness to new pipe-grade PE materials. A further complication is the extent of stress redistribution that occurs along the specimen ligament as craze fibrils creep under constant load. For a given test geometry, this redistribution complicates the evaluation of the stress dependency of slow crack growth.

A rigorous assessment of several American PE resins has been previously conducted by Kanninen et al. (1990) in which craze lengths were compared to the size of the stress singularity region at the crack tip. Kanninen et al. (1990) used Finite Element computation to numerically determine the stress field ahead of the crack tip in three-point bend (3PB) specimens machined from the wall thickness of different PE pipes. By comparing this computed stress field to the singular stress field predicted by LEFM theory (Equation 6.11), a region of dominance could be defined. For example, for the particular 3PB specimens used, the computed stress field was within 90% of the LEFM singular stress field over 6% of the ligament length. Therefore, the criterion for LEFM applicability was that craze length should be less than 6% of the ligament length (Kanninen et al. 1990). Whilst LEFM conditions were found to apply for older PE materials, newer grades of PE exhibited craze lengths that exceeded this 6% threshold value. Following the terminology of Kanninen et al. (1990), the brittle PE materials, which can be categorized by LEFM are classified as category I and the more recent tough materials as category II. As noted by Kanninen, Kuhlman, and Mahmoun (1993), the LIFESPAN prediction tool developed for PE gas pipes should only strictly be applied to category I materials.

Whereas Kanninen et al. (1990) used Finite Element stress analysis to determine the applicability of LEFM theory to PE materials, the ratio between the material yield strength and its cohesive strength and a quantity defined as the *T*-stress can also be used Broberg (1999). According to Broberg (1999), the cohesive strength is the stress required to separate a "unit cell" of material, where the unit cell is the smallest material unit which contains sufficient information about the crack growth properties of the material. For a given crack size and structure geometry, the *T*-stress represents the stress that acts parallel to the crack plane (in contrast to *K* in Equation 6.10, which characterizes the stress perpendicular to the crack plane). As described by Wang (2002), the *T*-stress can substantially alter the level of crack tip stress triaxiality (the presence of tensile stress on three orthogonal axes) and influence crack tip constraint. For example, a

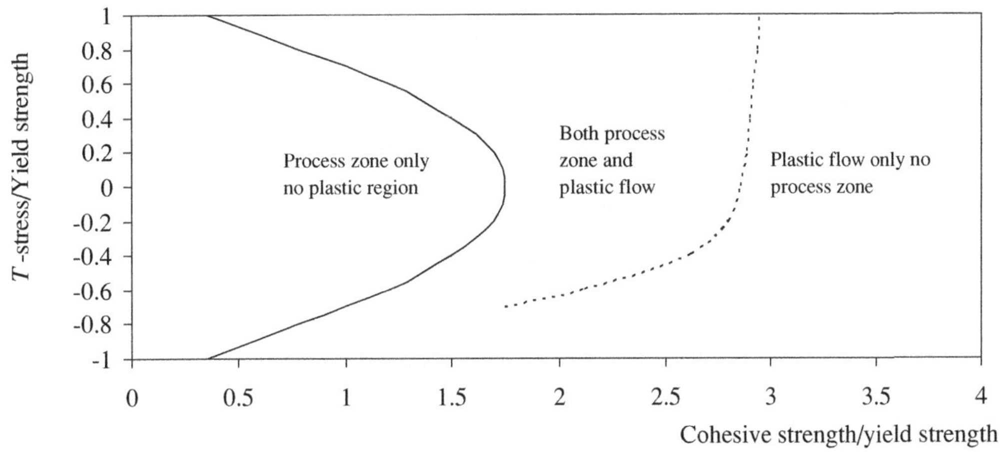

Figure 6.4 *T*-stress and cohesive strength/yield stress ratios with corresponding failure modes (adapted from Broberg, 1999)

positive *T*-stress strengthens the level of crack tip triaxiality and leads to high crack tip constraint (Wang, 2002). Under these conditions, brittle fracture rather than ductile yielding is promoted. In contrast, a negative *T*-stress reduces the level of crack tip triaxiality and leads to the loss of crack tip constraint (Wang, 2002). Under these conditions, ductile deformation rather than brittle fracture is promoted. Figure 6.4 illustrates how *T*-stress, material yield strength (σ_Y) and cohesive strength (σ_C) combine to promote different failure modes (Broberg, 1999).

As shown in Figure 6.4, for a given *T*-stress/σ_Y ratio, if the ratio σ_C/σ_Y is sufficiently low then a process zone develops in an elastic environment. For PE materials, this process zone will take the form of a crack tip craze and *provided that the craze length is less than 6% of the remaining un-cracked ligament thickness*, LEFM theory could be applied (Kanninen et al. 1990). In an intermediate case, a process zone develops within a plastic environment and finally, if σ_C/σ_Y is sufficiently high, plastic flow develops with no process zone and failure occurs by plastic collapse rather than fracture. In both these cases, LEFM should not be applied.

Therefore, for a given *T*-stress and provided that the material yield strength and cohesive strength are known, it is possible to determine the applicability of LEFM theory to different PE materials. As demonstrated by Duan and Williams (1998), the cohesive strength (or craze strength) can be measured approximately using circumferentially deep notched tensile specimens as discussed previously in chapter 5.

ELASTIC-PLASTIC FRACTURE MECHANICS (EPFM)

In the event that LEFM theory cannot be validated for a particular PE material, an alternative analysis method is required to predict lifetime. As discussed above, fracture in more recent generations of PE materials involves extensive plastic deformation and the elastic fracture analysis must be extended to incorporate additional energy dissipated during fracture (Williams, 2000).

This elastic-plastic regime is located in the middle region in Figure 6.4 and involves a number of additional complications that are not present in the elastic regime. Firstly, there is inherent non-linearity in material deformation and large geometry changes that accompany fracture.

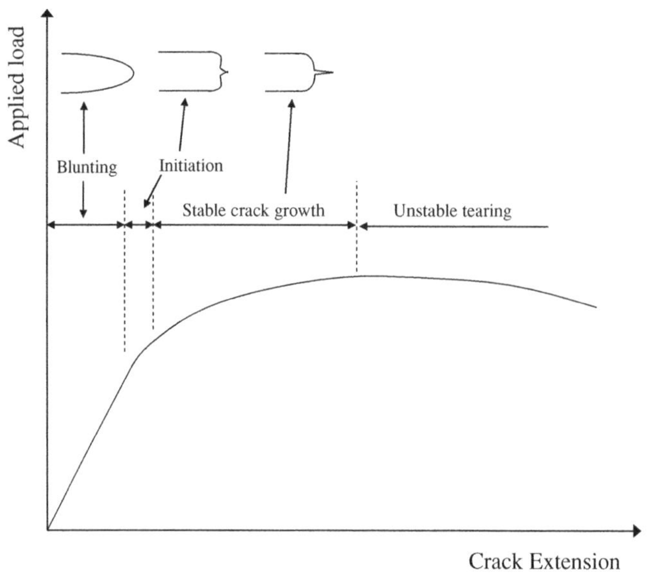

Figure 6.5 Schematic diagram illustrating the stages of ductile fracture (adapted from Kumar, German, and Shih, 1981)

A further complication is the significant crack blunting that occurs prior to crack initiation and the stable crack growth that can occur prior to unstable final fracture. This is in contrast to the brittle mode of fracture in which crack initiation and brittle fracture are essentially coincident (Kumar, German, and Shih, 1981). The various stages of ductile fracture are shown in the schematic diagram in Figure 6.5.

An engineering approach uses Elastic Plastic Fracture Mechanics (EPFM) to predict ductile fracture and is based on the crack tip characterizing parameter known as the J-integral. As outlined by Williams (2000), J is an energy release per unit area of crack face for a non-linear elastic solid. However, J is also used for materials where a non-linear load-displacement relationship is brought about by plastic (rather than elastic) deformation. While the J-integral provides an energy release rate (expressed in KJ/m^2), it also corresponds to a crack driving force per unit width of crack front (expressed in Newtons/metre). The General Electric (GE) Handbook (Kumar, German, and Shih, 1981) contains tabulated solutions for J, obtained using finite-element stress analysis for ductile materials. For example, the J-value for an internally pressurized cylinder with a radial at the inner surface, the corresponding elastic-plastic crack driving force (J) is expressed as (Kumar, German, and Shih, 1981)

$$J = \frac{K_{ITOT}^2}{E} + \alpha\varepsilon_0\sigma_Y c h_1\left\{\left(\frac{p}{p_0}\right)^{n+1}\right\} \tag{6.16}$$

In Equation 6.16, K_{ITOT} is the total applied SIF from LEFM theory, E is the material Young's modulus; c is the un-cracked ligament length; p is the applied internal pressure; σ_Y is the material yield strength and the parameters α, ε_0 and n are obtained from the stress-strain relationship for the material. The dimensionless EPFM function h_1 is tabulated for a range of crack lengths and pipe sizes in the GE Handbook (Kumar, German, and Shih, 1981).

Although equations are omitted for simplicity, the applied SIF K_{ITOT} and the dimensionless factor h_1 in Equation 6.16 both depend on crack length. To account for the crack tip craze zone, EPFM theory defines an "effective" crack length as the original crack length a plus an additional length ϕr_y, which is some fraction of the crack tip craze zone.

$$a_e = a + \phi r_y \tag{6.17}$$

As outlined by Bernal, Lopez Montenegro, and Frontini (2003), the additional crack length is given by

$$\phi = \frac{1}{1 + (p/p_0)^2}, \; r_y = \frac{1}{6\pi}\left[\frac{n-1}{n+1}\right]\left(\frac{K_{ITOT}}{\sigma_Y}\right)^2 \tag{6.18}$$

Although there is no EPFM-based expression to predict time–dependent crack growth directly, the effect of time can be incorporated indirectly through the change in craze zone length. As outlined by Atkins and Mai (1988), the time dependence of material yield strength σ_Y follows a power law in the form

$$\sigma_Y = \sigma_0 t^{-q} \tag{6.19}$$

As time progresses, σ_Y decreases as per Equation 6.19 and consequently, the length of the craze zone will increase following Equation 6.18. According to Equation 6.17, this will also produce an increase in the effective crack length a_e. Finally, (although the equations are omitted for simplicity), this increase in effective crack length will produce corresponding increases in the applied SIF K_{ITOT} and the dimensionless EPFM function h_1. Finally, referring back to Equation 6.16 the net effect of the time dependent decrease in yield strength is to increase the crack driving force J.

Having defined the crack driving force using the J parameter, a failure criterion for ductile fracture is required. J-crack growth resistance values (often denoted as "J_R") can be determined for a particular material by following an experimental protocol described by the European Structural Integrity Society (ESIS) (Hale and Ramsteiner, 2000). The recommended test method uses either compact tension or single edge notch bend (SENB) specimens, which are subject to a constant displacement rate until failure. For each specimen, J_R is determined by numerically integrating under the load vs. displacement plot resulting from each test. To fully characterize J_R for a particular material, a series of specimens are loaded to different final displacements and the amount of crack extension that occurs during the test is measured. By plotting crack growth Δa against the corresponding J_R value, a material J-resistance (J-R) curve can be generated. Previous tests on ductile polymers indicates that material J-R curves follow a power law form given by

$$J_R = A\Delta a^N \tag{6.20}$$

where A and N are empirical constants determined by fitting experimental J_R vs. Δa data. A critical J_R value required for crack initiation is arbitrarily defined as $J_{0.2}$, or the value of J_R that corresponds to 0.2mm of crack growth (Hale and Ramsteiner, 2000). As prescribed in the ESIS technical committee protocol, this corresponds to the material elastic-plastic crack initiation toughness

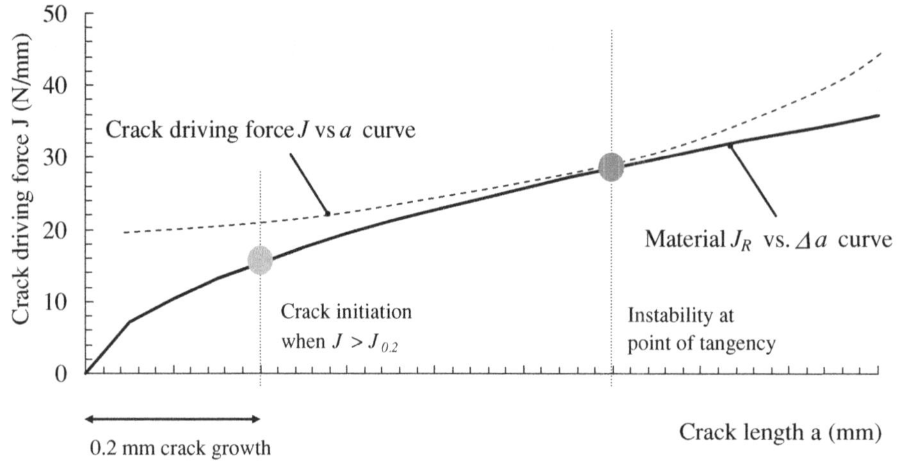

Figure 6.6 *J*-based failure criteria for elastic plastic fracture

J_C. Typical values for PE pipe materials are $A = 38.04$ and $N = 0.56$ (Bernal, Lopez Montenegro, and Frontini, 2003). Substituting these values into Equation 6.20 for a crack extension Δa = 0.2mm, gives an elastic-plastic crack initiation toughness of J_C = 15.45 N/mm. In the elastic-plastic regime, crack initiation is assumed to occur when the crack driving force per unit width J exceeds J_C. After crack initiation, failure can occur by two possible mechanisms. Firstly, stable crack growth through the pipe wall can occur when the effective crack length calculated from Equation 6.17 exceeds the pipe wall thickness, b. Secondly, unstable crack growth can occur after some amount of stable crack growth in ductile materials. After crack initiation, unstable crack growth is predicted to occur at the point of tangency between the applied J vs. a curve (calculated from Equation 6.16 and the material J_R vs. Δa curve. At this point, the gradient of the two curves is equal and the amount of stable crack growth before instability can be determined. The criterion for crack initiation and subsequent failure in the elastic-plastic regime are written as

$$\text{Crack initiation: } J = J_{0.2} = J_C \tag{6.21}$$

$$\text{Stable crack growth through pipe wall: } a_e = b \tag{6.22}$$

$$\text{Crack instability: } \begin{cases} J = J_R \\ \dfrac{\partial J}{\partial a} = \dfrac{dJ_R}{da} \end{cases} \tag{6.23}$$

The criteria for initiation and unstable ductile fracture are shown graphically in the schematic diagram in Figure 6.6.

The Applicability of EPFM Theory

While EPFM theory accounts for plastic deformation that accompanies ductile fracture, actual crack growth is required to occur for material *J-R* curves to be physically meaningful. As demonstrated by Bernal, Lopez Montenegro, and Frontini (2003), some tough PE materials form large crazes and exhibit extensive crack blunting, which appears as apparent crack extension in laboratory tests but merely represents plastic deformation and collapse at the crack tip. In this fully plastic regime, the use of a single parameter fracture criterion (such as the *J*-integral) would be valid only if the cracked specimen exhibited a relatively high level of stress triaxiality and crack tip constraint (Bernal, Lopez Montenegro, and Frontini 2003). In the absence of such constraint, critical fracture resistance values would exhibit a size and geometry dependence, and material *J-R* curves (determined from coupon-sized specimens) would not be transferable to other geometries such as buried pipes. If crack blunting occurs instead of actual crack growth, the material *J-R* curve should be linear rather than follow a power law. This linearity provides a check for the validity of EPFM.

If EPFM theory is not applicable, the problem then reduces to one in which the main effect of the crack is to reduce the cross-section of the structure. As described by Bernal, Lopez Montenegro, and Frontini (2003), ultimate failure is predicted to occur by plastic collapse of the remaining un-cracked ligament when the applied stress exceeds the material yield strength. To assess the applicability of both EPFM and limit load analysis methods, Bernal, Lopez Montenegro, and Frontini (2003) compared failure predictions for pre-cracked PE pipes under different loading configurations. Using axially-notched PE pipes under internal pressure, Bernal, Lopez Montenegro, and Frontini (2003) demonstrated that critical pressures required for failure could be best predicted using an EPFM *J*-based failure criterion. However, for circumferentially cracked PE pipes under remote axial tensile loads, the most accurate predictions of critical failure loads were provided by a simpler limit load analysis.

CRAZE MECHANICS-BASED PREDICTION

Kanninen et al. (1990) proposed that for those PE materials that contravene LEFM theory, detailed models of the crack tip craze are required to assess long-term performance. To this end, Duan and Williams (1998) attempted to predict the time to crack initiation from experimentally determined long-term craze-strength data. While craze strength data can be obtained using circumferentially deep-notched tensile specimens, the problem is to relate the uniform tensile stress state that exists in CDNT specimens, to the actual stress state in the craze zone that is formed in other component geometries.

From observations of plane strain single-edge notch bend (SENB) tests on tough PE materials, Duan and Williams (1998) proposed that upon initiation, a crack jump was observed through the crack tip craze zone. For this sudden failure of the entire craze zone to occur, it was reasoned that two conditions must be met. Firstly, the stress must be constant along the entire craze zone and secondly, the load bearing time for different points along the craze must be the same at failure. It was proposed that these conditions can only be met if the stress redistribution time due to craze fibril creep is much shorter than the time to crack initiation. This was observed in long-term SENB tests, where the time for the craze to reach 85% of its final length at failure was less than one tenth of the total time to crack initiation (Duan and Williams, 1998). It was therefore proposed that the stress redistribution time in tough PE materials occupies only a small portion of the total fracture initiation time. As outlined by Duan and Williams (1998), this implies that after this short stress redistribution period, a uniform tensile stress exists in the craze zone of *any* specimen geometry which is the same as the tensile stress in the circumferentially deep notched tensile test. Therefore, provided the tensile stress in the craze zone is know for the component under investigation (in this case and SENB specimen), the time to crack initiation could be predicted using measured craze strength data.

For a given component geometry, the actual stress in a craze zone after creep induced stress redistribution can be estimated using the concept of a "reference stress." According to Duan and Williams (1998), the reference stress concept originated from the creep analysis of metal rectangular beams under uniform bending moments. An analysis conducted by Schulte (1960) revealed the existence of a particular location in the beam where the stress remained invariant during stress redistribution by creep. It was postulated that this value of the invariant stress could serve as a reference stress for the whole beam and any uni-axial creep and rupture tests conducted at the reference stress could reproduce the behavior of the beam (Duan and Williams, 1998). According to Duan and Williams (1998), an estimate of the upper bound of the reference stress (σ_{ref}) can be derived as

$$\sigma_{ref} = \frac{P\sigma_Y}{P_U} \tag{6.24}$$

where P is the current load on the component, P_U is the rigid-plastic collapse load for the component and σ_Y is the yield stress. Following the reasoning outlined above, the time dependence of reference stress is the same as the time dependence of craze stress determined using circumferentially deep notched tensile specimens under constant tensile load. For one particular pipe-grade PE material this time-dependency was found to be

$$t_i = \left(\frac{\sigma_{ref}}{21.45}\right)^{-20} \tag{6.25}$$

where t_i is the initiation time (in hours) (Duan and Williams, 1998). Therefore, by using Equation 6.24 to calculate reference stresses for SENB specimens, crack initiation times could be predicted using Equation 6.25. Using the reference stress concept, reasonable agreement has been obtained between predicted and measured initiation times in SENB specimens under constant applied load (Duan and Williams, 1998).

The Applicability of Craze Mechanics-Based Predictions

While this technique could potentially predict crack initiation in buried pipes based on independently measured craze properties, reference stress solutions for crack tip crazes in buried pipes are required. Also, the subsequent stages of crack growth and/or ductile tearing (as described by EPFM theory) are not accounted for using craze mechanics.

RELIABILITY-BASED APPROACHES

In addition to predictive models for crack growth and failure, service life prediction tools for PE gas pipes have also been developed using reliability-based approaches. Whereas the approaches presented above provide a single-valued (deterministic) prediction of PE pipes service lifetime, a reliability approach includes the inherent uncertainty in the failure process. In doing so, service lifetime is represented by a probability distribution (with a mean value and a variance).

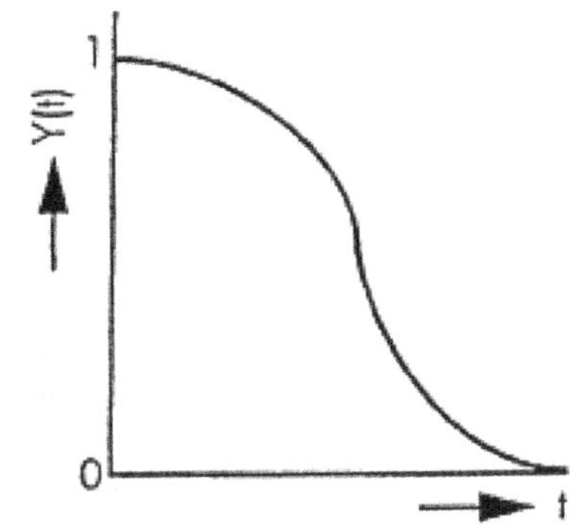

Figure 6.7 Survival curve (adapted from van Driel, 1997)

The benefit if such an approach is that probabilities of failure can be defined which are useful for pipeline risk assessment. Alternatively, failure rates can be estimated, which allow future investment needs to be forecast based on the expected cost of pipe repair or replacement.

Using a reliability-based approach, GASTEC NV in The Netherlands developed a system to assess quality and estimate remaining lifetime of buried gas pipelines. As described by van Driel (1997), the GASTEC NV methodology is based on the general statistical background of human lifetime modeling as typically used by insurance companies. The theory is based on a "survival curve," as shown schematically in Figure 6.7.

This curve represents the probability of a particular unit (i.e., a pipe) surviving as a function of accumulated lifetime. In general, the probability of unit survival decreases with increasing lifetime. However, as noted by van Driel (1997), the exact position of the curve and the slope at different lifetimes depends on the influence of the surrounding environment. According to van Driel (1997), the survival curve in Figure 6.7 can be characterized by the so-called Weibull probability distribution function given by

$$Y(t) = \exp-\left[\left(\frac{t}{\alpha}\right)^{\beta}\right] \tag{6.26}$$

where Y represents the probability of survival, t is the number of time units, α is the so-called characteristic age and β describes the shape of the curve. These parameters that describe the lifetime probability distribution are also dependent on the operating and installation environment surrounding a particular pipeline. Variables such as pipe operating pressure and external load may be expected to influence α and β.

Within a particular time interval, the probability of failure occurring can be calculated using the probability density function. Using the notation of van Driel (1997), this is given by

$$F = \beta\alpha^{-1}\left(\frac{t}{\alpha}\right)\exp-\left[\left(\frac{t}{\alpha}\right)^{\beta}\right] \tag{6.27}$$

which results in a curve as shown schematically in Figure 6.8.

According to van Driel (1997), the number of failures for a certain population as a function of time is given the intensity of failure, z. Again using the notation of van Driel (1997)

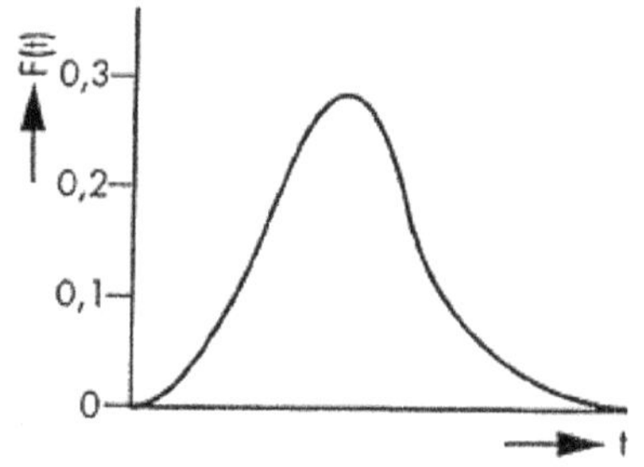

Figure 6.8 Chance of failure curve (adapted from van Driel, 1997)

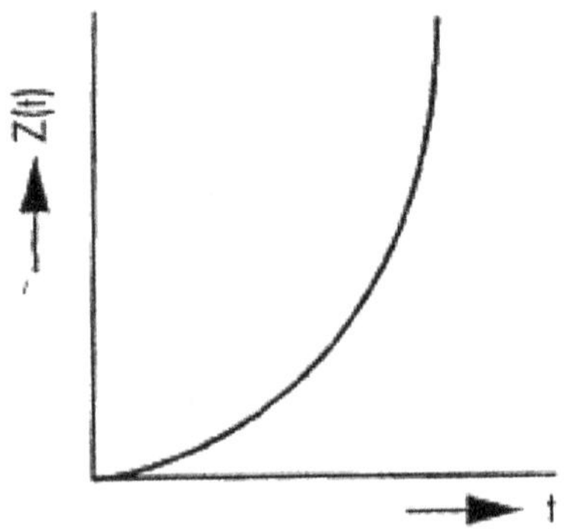

Figure 6.9 Failure rate curve (adapted from van Driel, 1997)

$$z = \frac{1}{Y}\frac{dY}{dt} = \beta\alpha^{-1}\left(\frac{t}{\alpha}\right)^{\beta - 1} \tag{6.28}$$

z represents a failure rate (or percentage of pipe length that has failed) and follows a time dependence of the type shown schematically in Figure 6.9 (van Driel, 1997).

Based on this underlying reliability theory, the GASTEC NV methodology attempted to quantify α and β parameters that describe the Weibull probability distribution for PE pipe lifetime in a range of operating and installation conditions. As outlined by van Driel (1997), an inventory study was conducted on available information from research and maintenance records. This provided a list of the most important factors that influence degradation and failure of a pipe in service. An inventory of available information from records at gas companies was then conducted to determine the maximum level of detail for input data. According to van Driel (1997), combining these inventories resulted in matrices of all influencing factors and their accompanying grades (levels of detail). The next step in the process was to define the values for the difference influencing factors. These values were based on analysis of results from existing research, such as that conducted by GASTEC on the influence of soil loads on buried PE pipes. Although no further information is given (van Driel, 1997), it is assumed that the reduction in lifetime due to different combinations of influencing factors was then used to construct a sample of pipe lifetimes. This sample could then be fitted to Equation 6.26 to estimate parameters α and β which describe the Weibull distribution for pipe lifetime.

The Applicability of Reliability-Based Models

A reliability-based approach to service life modeling is attractive since it should account for the inherent uncertainty in the pipe failure process. The approach adopted by GASTEC NV (van Driel, 1997) attempts to account for uncertainty in service lifetime through the Weibull probability distribution function. However, the main difficulty is to assign physically meaningful values to the parameters (in this case α and β) that describe the distribution function. It is not clear from previous work (van Driel 1997) how these values are obtained and also if they have a physical basis. For example, it is reasonable to expect that they would depend at least in part on the slow crack growth resistance properties of the PE pipe material If these difficulties can be resolved, a potentially useful approach may be to combine service life predictions from a physically-based model (using LEFM theory for example) with the reliability model presented above. As outlined by van Driel (1997), estimated future failure rates in pipe networks can allow future investment needs to be identified which in turn can allow utilities to plan accordingly to meet them.

CONCLUDING REMARKS

This chapter has presented several methods for long term failure prediction in pipe-grade PE materials. While each is applicable to particular materials, no single approach can be confidently applied to all PE materials.

In general, approaches to failure prediction can be divided into empirical and theoretical methods. Empirical methods rely on experimental data from fracture property assessments or hydraulic pressure testing at different applied stresses and temperatures. Based on this data, failure times from elevated temperature testing (which can be obtained in practical timeframes) can be transformed to give corresponding failure times at ambient temperature. The so-called Rate Process Method and Bi-Directional Shifting Method are examples of these techniques that have been developed through research into PE gas pipe lifetimes. While these techniques are useful, evidence in the literature suggests that the mechanism of crack growth that is observed at elevated temperatures differs from that which occurs at ambient temperature. However, sufficient correlation exists to rank the performance of different PE materials and provide qualitative correlations with observed field performance. For example, those materials that exhibit extended failure times from RPM testing generally exhibit low failure rates in service. One issue with empirical approaches is the increased testing time that is required for new PE materials with improved slow crack growth resistance. Even at elevated temperatures, the performance of these materials is such that test times are impractical.

An alternative method is to develop a lifetime prediction model based on Linear Elastic Fracture Mechanics (LEFM) theory. LEFM uses the concept of a single parameter known as the applied Stress Intensity Factor (SIF) that characterizes the stress field at the tip of a crack in a PE pipe. The dependence of crack growth rate on the applied SIF can be determined using small laboratory scale tests on coupon samples and then transferred through a geometrical correction factor to the geometry of a flawed pipe in service. While this approach can be applied to older PE materials, new PE materials can form a large craze zone at the crack tip which renders LEFM invalid. An extension to LEFM theory is Elastic Plastic Fracture Mechanics (EPFM), which can be used to model post-yield fracture that occurs after plastic deformation in ductile materials. However, care must be taken to distinguish plastic fracture failure from crack tip blunting. In the latter failure mechanism, EPFM failure predictions cannot be transferred from small test geometries to larger pipes in service.

Recent advances in failure mechanism research have led to a novel test method which isolates the craze formation and separation process in PE materials. The time dependence of craze failure stress obtained under static tensile load can be used to predict the time to crack initiation in other test geometries such as single edge notch bend specimens. While this approach is promising, it has yet to be applied to practical cases such as a flawed pipe under in service operating loads.

In addition to empirical and mechanism-based failure models, reliability-based approaches have also been proposed, which account for uncertainty in the failure process. In this method, the service lifetime of a PE pipe is represented by a suitable Probability Distribution Function (such as the Weibull Extreme Value distribution). The approach is attractive since it should estimates the probability of pipe failure, which in turn can be used to estimate average failure rates in pipe networks or conduct risk assessments. Both of these performance measures are useful for pipe asset management. However, the main difficulty is to assign physically meaningful values to the parameters that describe the lifetime distribution function. It is not clear from previous work (van Driel 1997) how these values are obtained and also if they have a physical basis. If these difficulties can be resolved, a potentially useful approach may be to combine service life predictions from a physically-based model (using LEFM theory for example) with the reliability model.

CHAPTER 7
FRACTURE PROPERTIES OF PIPE GRADE PE MATERIALS

INTRODUCTION

In chapter 6, modeling approaches used to predict long-term performance of PE pipes were discussed. While tests such as the Rate Process Method are very useful tools to rank the performance of the different PE materials in pipe form, important performance-related information can also be obtained from tests on small coupon samples. Carefully selected fracture-type tests can not only rank different PE materials in the same order as pipe tests, but also provide essential input information to assist the development of a service life prediction model for PE pipes failure. To this end, this chapter presents test methods used to determine fracture properties and results from different PE materials from Australia and the US. Comparing these properties gives an indication of those pipes that are more likely to fail by fracture in the field.

Previous research has shown that the formation and ultimate separation of the crack tip craze is crucial to the slow crack growth resistance of PE materials. The first goal of this chapter is to use an experimental method which isolates the craze formation and separation process and measure "craze performance" for different materials. The second goal of this chapter is to determine which (if any) of the PE materials conform to single parameter Fracture Mechanics theory. If single parameter fracture mechanics can be applied, crack growth and fracture could be characterized by parameters such as the applied Stress Intensity factor (in the case of linear elastic behavior) or the *J*-crack growth resistance (for elastic plastic fracture). The advantage is that the dependence of crack growth rate on these parameters can be measured on using small coupon samples in the laboratory and then transferred to a realistic case of a pipe in service through a geometric correction factor. Checking the validity of single parameter fracture mechanics is an essential first step before developing a failure model.

CRAZE PROPERTY TESTING

As described in chapter 5, the circumferentially deep-notched tensile (CDNT) test can be used to isolate and examine the important craze zone that forms at the tip of a propagating crack in PE materials. Rectangular bar specimens are used with a symmetrical notch, machined around the circumference at the mid point and then razor-notched. The deep circumferential notch develops a highly constrained region in the small remaining ligament. As a consequence of this geometrical discontinuity, a tri-axial stress state and high magnitude of local stress is achieved (Pandya and Williams, 2000). Therefore, when loaded in tension, the high constraint promotes craze growth in the ligament, which is the most severe damage mechanism likely to occur for a flawed pipe in service. Figure 7.1 shows a photograph of a CDNT test in progress.

To assess craze behavior under constant displacement rate, CNDT specimens were tested in a screw driven Instron machine in air at 20°C. Loads were recorded using a 10 kN load cell and converted to a craze stress by dividing by the ligament area. Local separation of the craze was measured directly using a clip gauge extensometer as shown in Figure 7.1. To assess craze behaviour under constant load conditions, specimens were mounted in grips and connected to a pin jointed lever arm. A mass was the applied at the end of the lever arm to apply the required static

Figure 7.1 Circumferentially deep-notched tensile (CDNT) specimen

Table 7.1
PE material grades assessed using the CDNT test for craze behavior

Sample	Origin	PE grade	Relevant Standard	Hydrostatic strength category (or Hydrostatic Design Basis) at 50-year intercept
PE80C (AUS)	Australia	PE80C	AS 4131	8.0 MPa (1,160 psi)
PE80B (AUS)	Australia	PE80B	AS 4131	8.0 MPa (1,160 psi)
PE100 (AUS)	Australia	PE100	AS 4131	10.0 MPa (1,450 psi)
PE3408 (US)	USA	PE3408	AWWA C906	11.0 MPa (1,600 psi)

tensile stress to the ligament. As with constant displacement rate tests, local craze extension was measured using a clip gauge extensometer.

To assess the craze strength properties of PE materials, CDNT specimens were produced by compression molding pellets of raw material into rectangular plaques, (195mm × 135mm × 20mm thick). A rectangular mould was designed to hold the required volume of raw material and allow for excess molten material to be forced out through channels under compression. The mould was compressed between two heating platens up to a temperature of 200°C and an applied platen pressure of 1 MPa. After compression for a period of 3 hours to ensure a uniform temperature distribution throughout the molten polymer, the mold was cooled rapidly to room temperature by passing coolant through channels in the heating platens. Rectangular bar specimens (16mm × 16mm × 120mm long) were then cut from each compression molded plaque. Following the procedure outlined by Pandya and Williams (2000), a circumferential notch was introduced with a single point cutting tool of tip radius ≤20 μm by rotating the specimen in a four jaw self centering chuck on a lathe at low rotation speed. The notch tip was sharpened, also on a lathe such that the final ligament/bulk area ratio was 1:10 (Pandya and Williams, 2000). Table 7.1 gives details of the different the PE materials assessed.

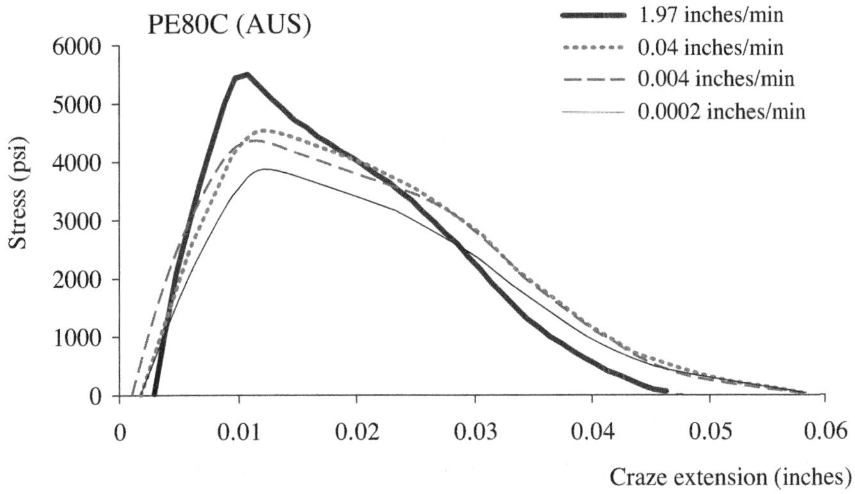

Figure 7.2 Traction-separation curves for PE80C (AUS) at constant displacement rate in the CDNT test

It should be noted that no further information concerning the molecular morphology and hence resistance to slow crack growth is known for these samples. As reported previously, these properties can vary within a particular PE material designation. Therefore, the properties measured in this chapter apply to this particular set of samples only and should not be assumed as typical for all materials of that particular designation.

Craze Energy Absorption and Extension at Constant Displacement Rate

The load displacement traces in Figure 7.2 compare the traction separation behavior for PE80C (AUS) at a range of different constant displacement rates.

As shown, in each case the crazed stress increases to a maximum, which corresponds to the formation of craze fibrils as described in chapter 3. Craze stress then decreases with increasing ligament separation. This corresponds to craze growth through fibril extension until ultimate separation of the craze zone occurs at failure. In general, as cross head speed increases, the peak craze stress also increases, but ultimate craze extension decreases. As shown in Figures 7.3 to 7.5, this trend is also observed in the other PE material grades.

Since a comparison of individual traction-separation curves for each material at each test speed is difficult, alternative measures of craze behavior are used. For each material, craze performance can be characterized by three parameters:

- Peak craze stress or cohesive strength (σ_C)
- Post-peak extension to failure (δ_{max})
- Energy absorbed during craze separation (Γ)

The influence of displacement rate on each of these properties is compared for all PE materials in Figures 7.6 to 7.8. Note that for PE3408, the test at 0.4 inches/min cross head speed was invalid and an additional test was conducted at a cross head speed of 0.002 inches/min.

As shown, craze strength in all PE materials increases with increasing test speed, which indicates the visco-elastic nature of these polymers. However, the influence of test speed on craze

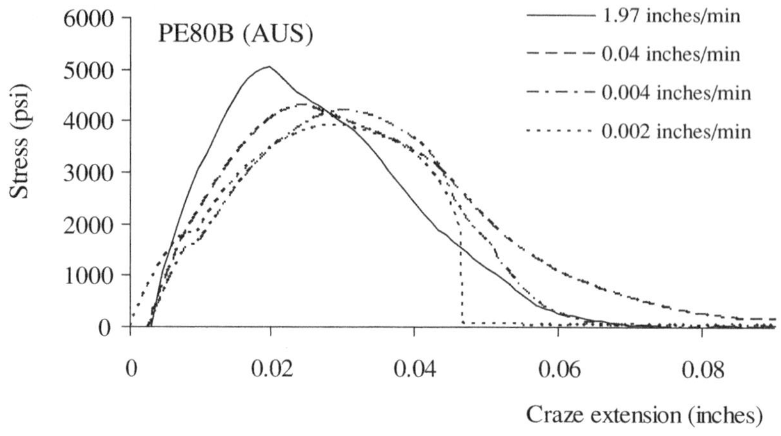

Figure 7.3 Traction-separation curves for PE80B (AUS) at constant displacement rate in the CDNT test

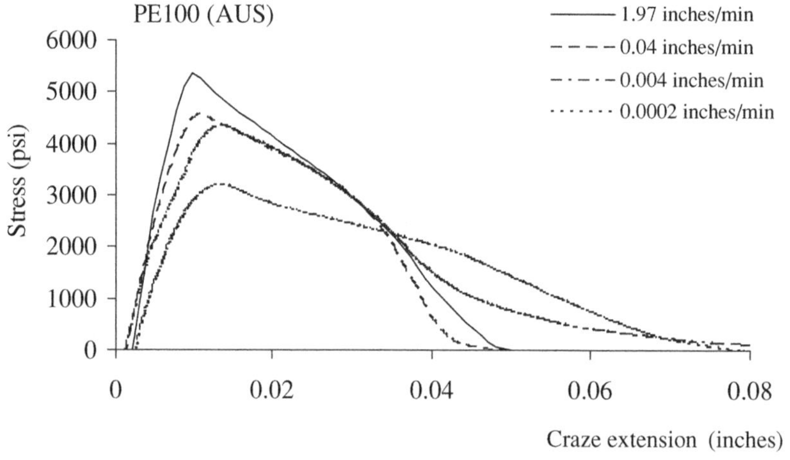

Figure 7.4 Traction-separation curves for PE100 (AUS) at constant displacement rate in the CDNT test

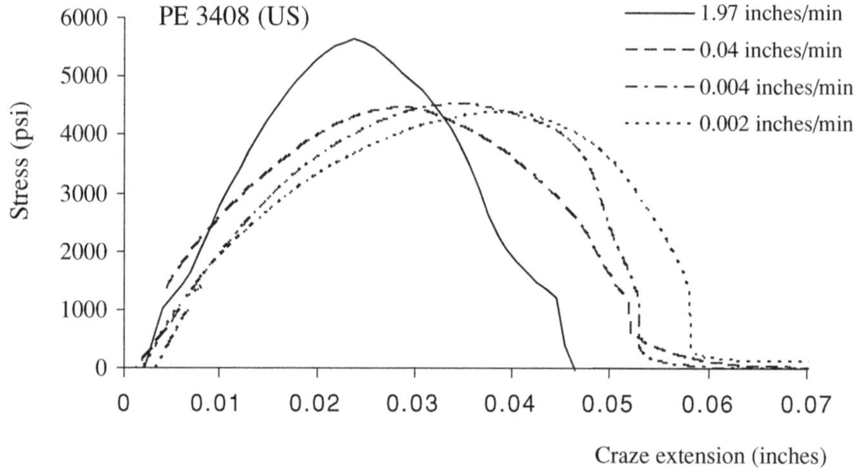

Figure 7.5 Traction-separation curves for PE3408 (US) at constant displacement rate in the CDNT test

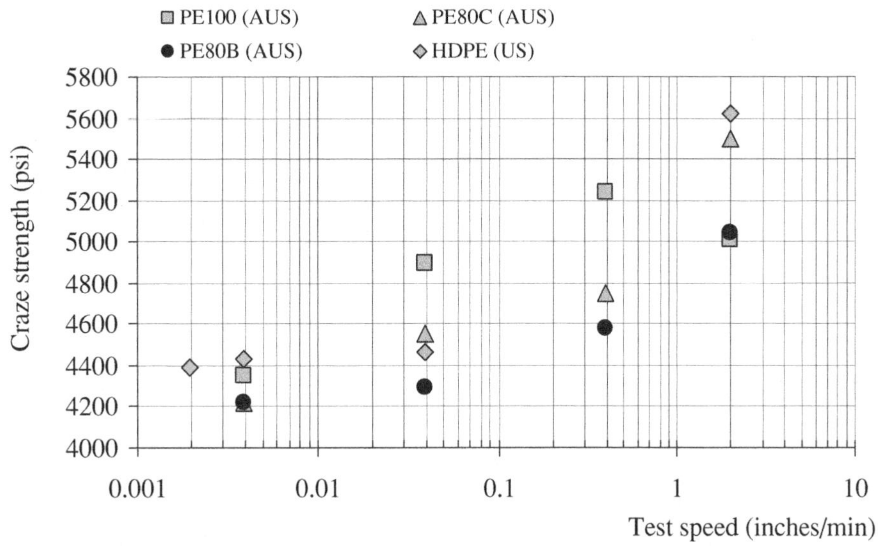

Figure 7.6 Effect of displacement rate on craze strength σ_C for different PE materials

strength differs for each material. In general, PE3408 (US) exhibits the highest craze strength at the lowest and highest test speeds and PE100 (AUS) exhibits the highest craze strength in the intermediate test speed range. PE80B (AUS) consistently exhibits the lowest craze strength across the range of test speeds.

In comparison to Figure 7.6, the ranking of materials in terms of post peak craze extension is generally reversed. Although the lowest craze strengths were observed in PE80B (AUS), this material exhibits the largest craze extensions. In contrast, the lowest craze extensions to failure are observed in PE3408 (US) across the range of test speeds. It is also interesting to note that the dependence of craze extension on test speed differs between materials. For example, both PE80C (AUS) and PE100 (AUS) show a decrease in craze extension with increasing test speed. It is possible that at higher rates of deformation, the failure of the craze zone in PE materials may be influenced by adiabatic heating as plastic work is dissipated as heat. However, while local heating may occur during craze deformation, it has been proposed that localized melting is required to cause craze failure (Leevers, 1995). While the melting point could be exceeded under high rates of loading (such as impact) it is less likely to be exceeded under the moderate test speeds used in the current study. Furthermore, localized heating would be expected to act in all PE materials and, as shown in Figure 7.7, this is not the case. In PE80B (AUS), craze extension actually increases with increasing test speed.

Perhaps the most useful craze parameter is the total energy absorbed during craze separation per unit ligament area (Γ), which is calculated by integrating under the traction separation curves in Figures 7.2 to 7.5. This single parameter encapsulates the combined effect of craze stress and extension and provides the best indicator for craze performance. As shown in Figure 7.8, PE100 (AUS), PE80B (AUS) and PE3408 (US) exhibit high craze separation energies across the range of test speeds when compared to PE80C (AUS). Although PE3408 (US) exhibits relatively low craze extensions, the higher stresses required for this extension result in a greater amount of energy absorbed. In practice, this means that the relatively high energy absorption during craze formation and separation would confer a high crack growth resistance to these materials. It is interesting to note that PE3408 (US), PE80B (AUS) and PE100 (AUS), all exhibit an increase in Γ as

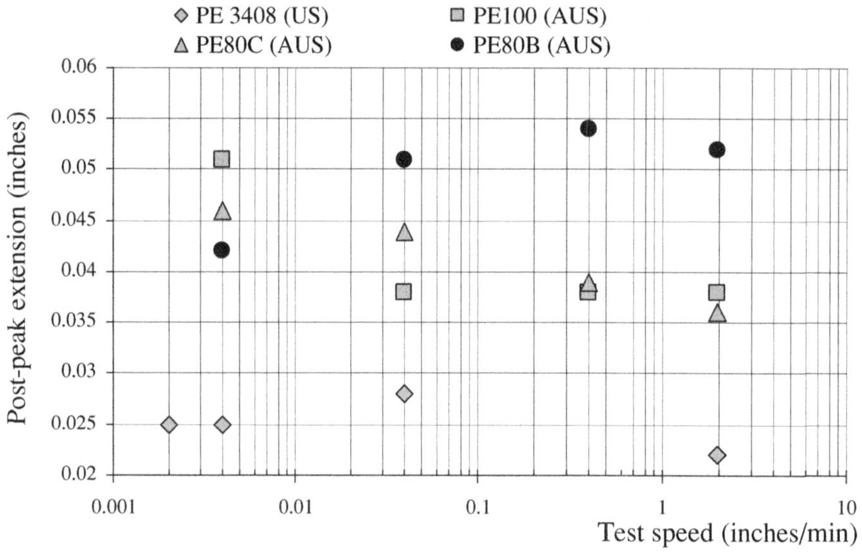

Figure 7.7 Effect of displacement rate on post-peak craze extension δ_{max} for different PE materials

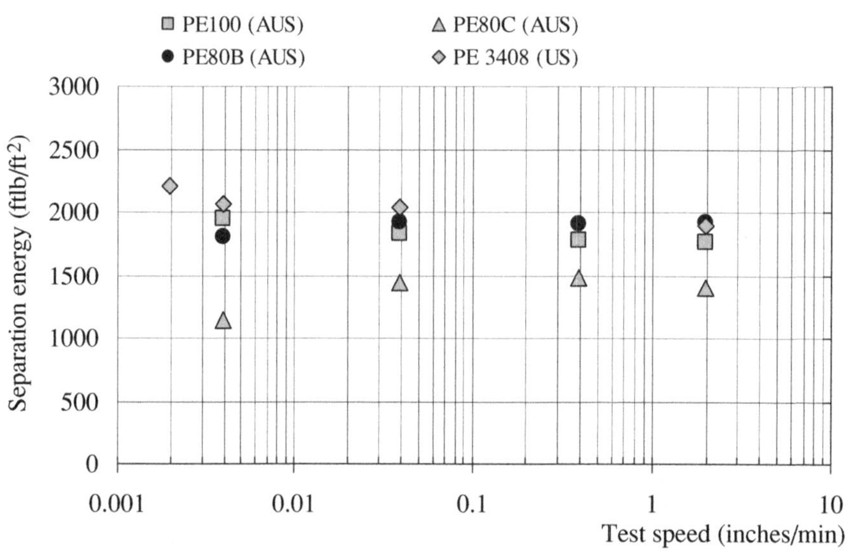

Figure 7.8 Effect of displacement rate on energy of craze separation Γ for different PE materials

test speed decreases. In contrast, Γ decreases with decreasing test speed for PE80C (AUS), which indicates more brittle behavior as reported by Pandya and Williams (2000).

Scanning Electron Microscopy (SEM) of Craze Failure

Further evidence of this different failure mode in PE80C (AUS) is clear from microscopic examination of the failed ligament. For example, Figure 7.9 shows a plan view of the failed ligament in PE80C (AUS) obtained using a Scanning Electron Microscope (SEM). A test speed of 0.004 inches/min was used in this particular case.

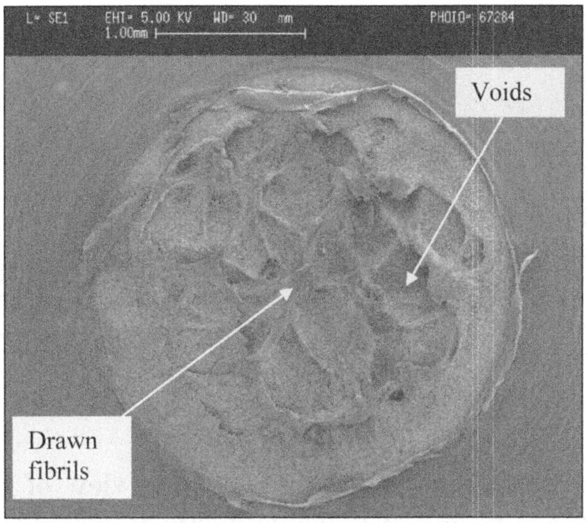

Figure 7.9 Scanning electron microscope image: Plan view of failed CDNT ligament in PE80C (AUS) at a displacement rate of 0.004 inches/min

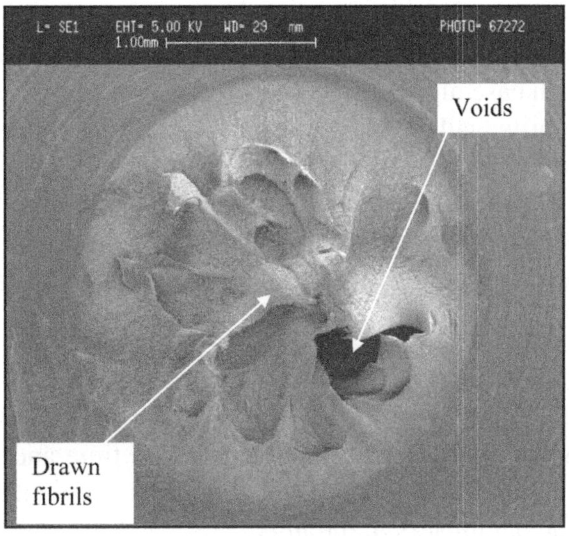

Figure 7.10 Scanning electron microscope image: Plan view of failed CDNT ligament in PE100 (AUS) at a displacement rate of 0.004 inches/min

As shown, the failed ligament is generally comprised of voided regions surrounded by fibrils of drawn material. For comparison, an equivalent image for PE100 (AUS) is shown in Figure 7.10.

In general, the ligament surface in PE100 (AUS) exhibits fewer voided regions and significantly larger fibrils, which indicates that the extent of material drawn from the bulk into craze fibrils is much greater in this case. It also interesting to note that the ring of drawn material around the ligament circumference in PE80C (AUS) is not present in PE100 (AUS). Again, this indicates that material was drawn from the bulk during craze extension without premature failure in PE100 (AUS). The greater extent of drawing in PE100 (AUS) is clear from the side view in each sample (Figure 7.11).

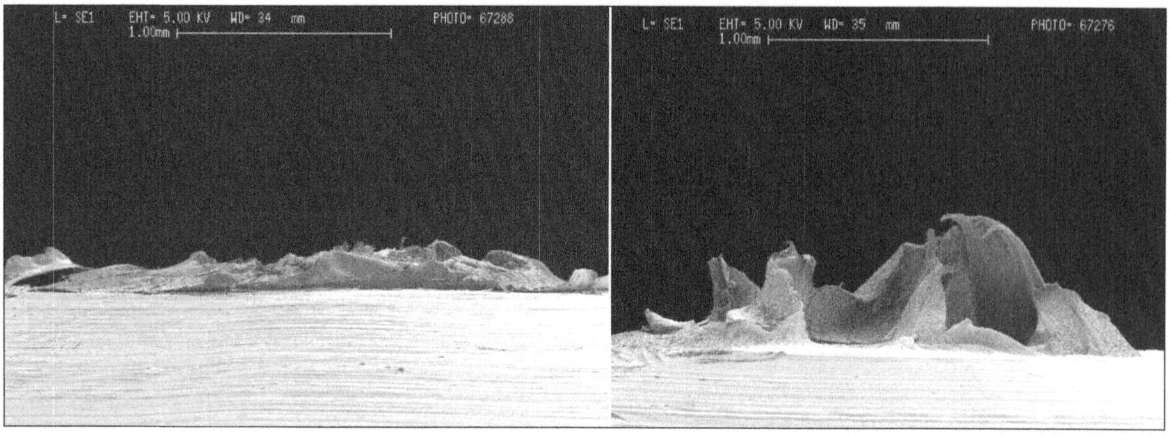

Figure 7.11 Scanning electron microscope image: Side view of failed CDNT ligament in PE80C (AUS) and PE100 (AUS) at a displacement rate of 0.004 inches/min

Time Dependent Craze Strength Under Static Tensile Load

To compare with the effect of displacement rate, craze tests were also conducted under static load conditions. Although energy of separation was not determined in these tests, time to failure was recorded in each case allowing the performance of each PE material to be ranked. For comparison with the currently available PE materials, CDNT samples were also prepared from a 1.5 inch diameter service connection that had failed in the US by point loading and fracture. Although the precise age of this sample was unknown, it is thought to be more than 10 years old. Figure 7.12 compares craze stress vs. failure time data for each sample.

While the data is clustered quite closely, an empirical power law in the form

$$\sigma_C = A t_f^{-m} \tag{7.1}$$

can be fitted, where σ_C is the applied stress, t_f is the failure time and A and m are material constants. As outlined by Duan and Williams (1998) a more convenient format for interpreting the physical implications of the parameters A and m is

$$t_f = \left(\frac{\sigma_C}{A}\right)^{-1/m} \tag{7.2}$$

where A is the stress level at which the material considered fails at unit time (hour) and can be termed the *general load bearing capacity parameter* (Duan and Williams, 1998). The index m, which scales the time dependence of material behavior is termed as the *damage evolution index* with $m = 0$ corresponding to time-independent fracture for which A corresponds to the critical fracture stress. With these definitions of parameters A and m, materials can be compared and graded in terms of long-term fracture resistance. A material with a higher A and lower m value will exhibit a better long term fracture resistance (Duan and Williams, 1998). Table 7.2 compares A and m parameters from CDNT tests under static loading conditions.

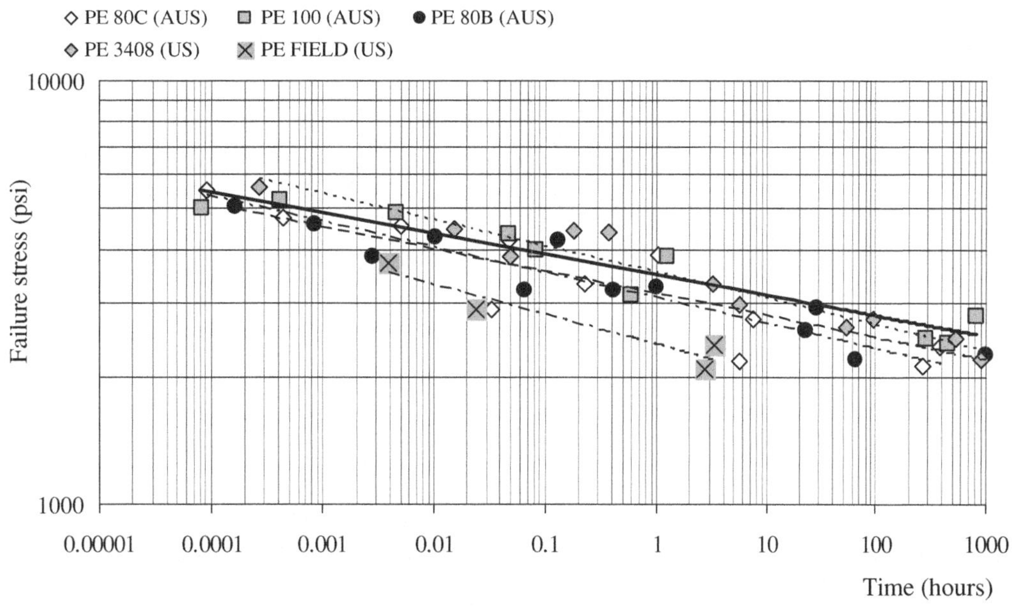

Figure 7.12 Craze stress σ_C vs. failure time under static load for different PE materials in the CDNT test

Table 7.2
Empirical long-term fracture resistance parameters from CDNT tests

Material	General load bearing capacity parameter *A* (psi)	Damage evolution index *m*
PE80C (AUS)	3,069.3	0.060
PE80B (AUS)	3,148.2	0.052
PE100 (AUS)	3,496.9	0.049
PE3408 (US)	3,531.9	0.062
PE FIELD (US)	2,383.9	0.071

PE3408 (US) exhibits the highest *A*-value, which corresponds to the highest short-term fracture resistance. However, it also exhibits a relatively high *m*-value indicating that damage will evolve at the highest rate in this material for a given applied stress (Duan and Williams, 1998). PE100 (AUS) exhibits the highest *m*-value indicating that damage will increase at the lowest rate. PE100 (AUS) also exhibits the second highest *A*-value which indicates that its overall long term fracture resistance is the highest of all materials tested using the CDNT method under static load. In comparison, PE80C (AUS) exhibits a lower *A*-value and a higher *m*-value, indicating a relatively low short-term fracture resistance and also a higher rate of increase in damage for a given applied stress.

As expected, the lowest performance of all samples was observed in PE FIELD (US). An initially low craze strength (corresponding to the lowest *A*-value) decreased much faster than the other samples tested (corresponding to the highest *m*-value). This is in good agreement with the known poor SCG resistance of the older PE material in comparison to currently available PE material grades (Mruk, 2006).

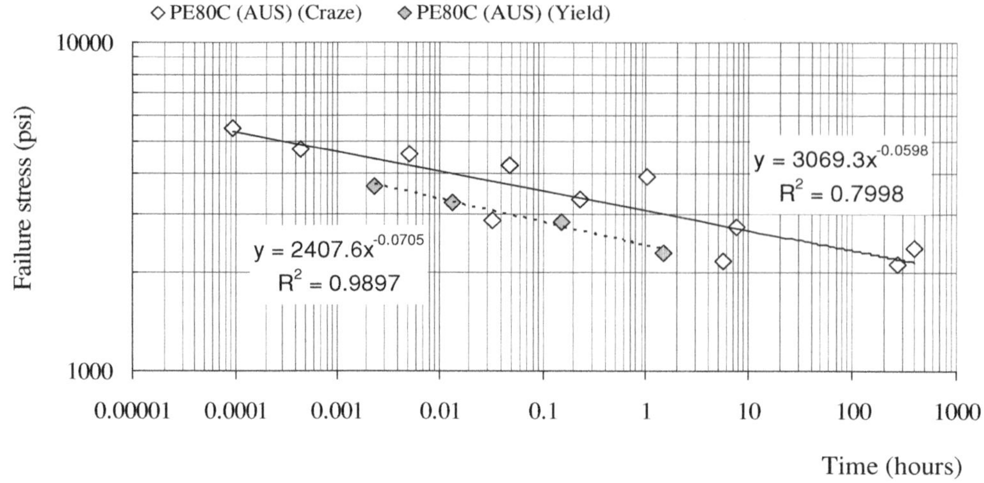

Figure 7.13 Comparison of craze stress and yield strength vs. failure time under static load for PE80C (AUS)

It should be noted that while Table 7.2 provides quantitative measures of material fracture resistance, this does not necessarily mean that linear elastic fracture mechanics (LEFM) can be applied to predict failure. In order to demonstrate the applicability (or lack of) of LEFM to these materials, measured craze strength must be compared to uniaxial yield strength data.

CRAZE STRENGTH-BASED ASSESSMENT OF LINEAR ELASTIC FRACTURE MECHANICS (LEFM) APPLICABILITY

As described in chapter 6, the ratio of measured craze strength (determined using the CDNT test) and the material uniaxial yield strength can be used to determine the type of fracture failure mode likely to occur in a particular PE material. To combine with results from the CDNT test, the uniaxial tensile strength of a number of different PE grades was determined in accordance with the Australian Standard AS 1145.2 Determination of tensile properties of plastics materials Part 2: Test conditions for molding and extrusion plastics (Standards Australia, 2001). As with CNDT tests, static load tests were also conducted using the pin-jointed lever arm described above.

The data in Figures 7.13 to 7.16 show comparisons between craze strength and yield strength for different failure times.

As shown in Figures 7.13 and 7.14 failure stress under yield and craze conditions show distinctly different time dependencies. As outlined by Duan and Williams (1998), this is attributed to the different modes of failure that occur in each test. Specimens in the uniaxial yield tests are under a state of plane stress with very little constraint in the through-thickness direction. This promotes failure by ductile yielding at a relatively low applied stress. In contrast, CDNT specimens in craze strength tests are exposed to a much higher level of stress triaxiality (and hence constraint) due to the presence of a circumferential notch. Higher applied stresses are required to overcome this constraint which restricts ductility and promotes crazing and a relatively brittle failure mode. In general, uniaxial yield strength decreases more rapidly with increasing time, compared to craze strength. This result was also observed by Duan and Williams (1998) and was attributed to the different rates and mechanisms of creep deformation occurring in each test. Due to increased constraint, material in the craze strength test contains a large number of voids which

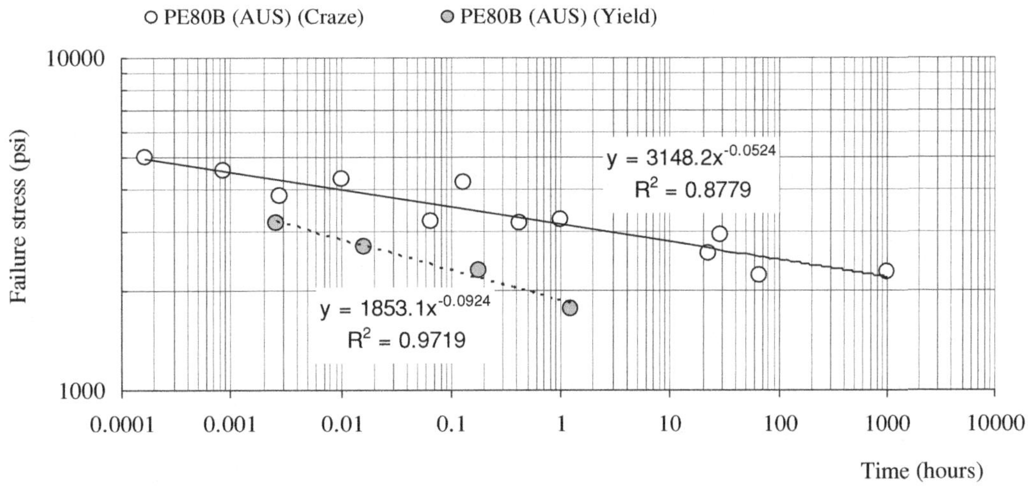

Figure 7.14 Comparison of craze stress and yield strength vs. failure time under static load for PE80B (AUS)

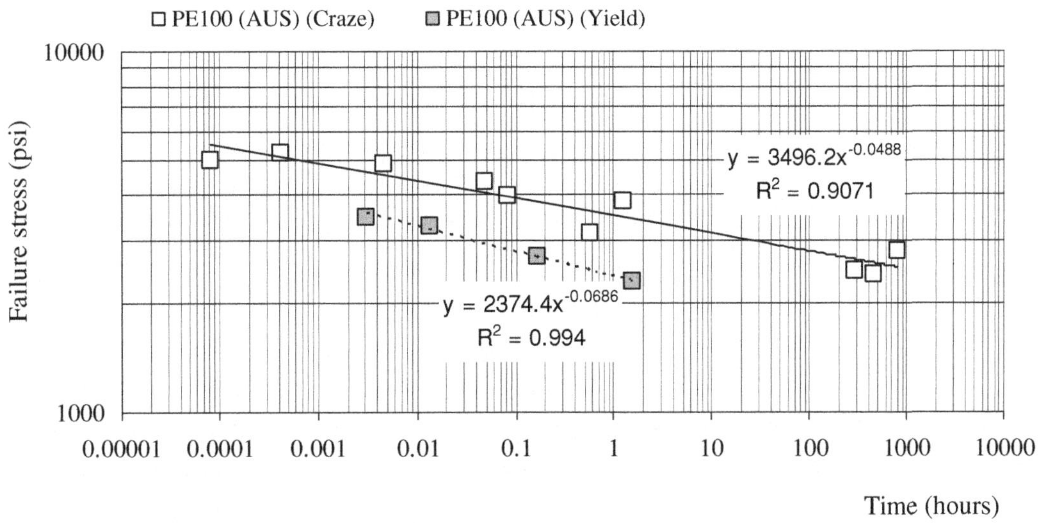

Figure 7.15 Comparison of craze stress and yield strength vs. failure time under static load for PE100 (AUS)

increase the rate of creep deformation compared to material in the uniaxial yield strength test. However, the craze extension obtained prior to failure is also greater than that required to yield material in the yield strength test. Therefore, the overall time to failure is generally greater in the craze strength test. However, it is interesting to note that for PE80C, the two curves are closer together compared to PE80C (AUS). This may correlate with the more brittle behavior exhibited by PE80C in Figures 7.8 and 7.12.

A similar trend to that observed in PE80B (AUS) is observed for PE100 (AUS) and PE3408 (US) materials.

As described previously in chapter 6, the ratio between material yield strength (σ_Y) and cohesive strength (σ_C) can be used to determine the deformation state that will exist at the crack

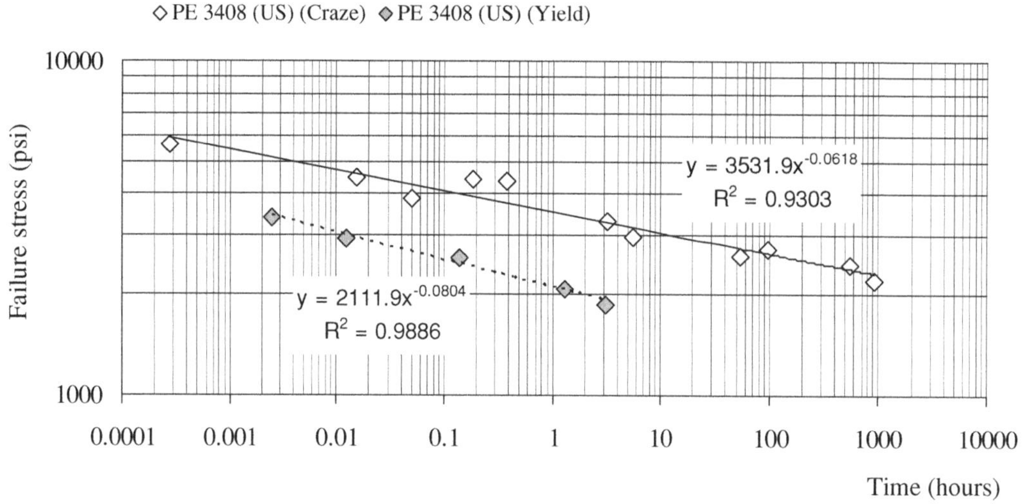

Figure 7.16 Comparison of craze stress and yield strength vs. failure time under static load for PE3408 (US)

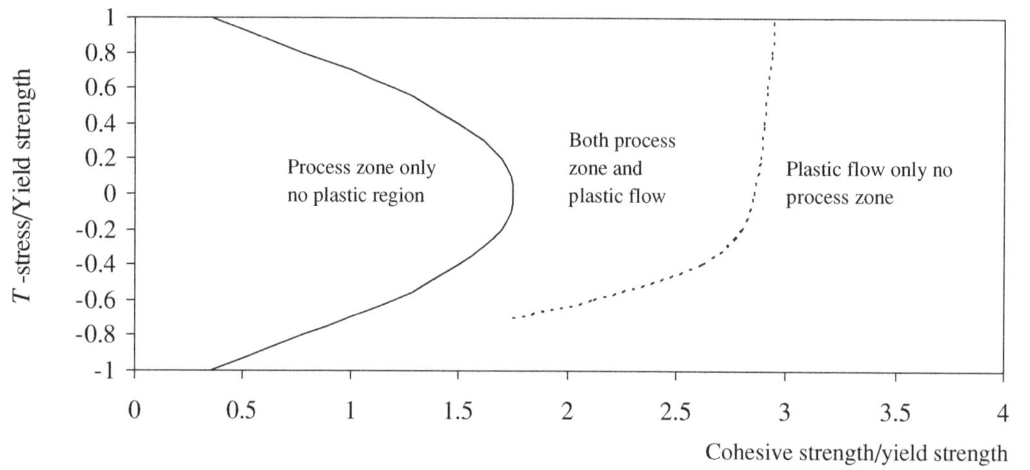

Figure 7.17 *T*-stress and cohesive strength/yield stress ratios with corresponding failure modes (adapted from Broberg, 1999)

tip ahead of a defect that is subjected to a given *T*-stress (Broberg, 1999). As described by Wang (2002), the *T*-stress represents the stress that acts parallel to the crack plane and influences crack tip constraint. A positive *T*-stress strengthens the level of crack tip triaxiality and leads to high crack tip constraint (Wang, 2002). Broberg's (1999) criteria for different crack tip deformation states in terms of σ_Y, σ_C and *T*-stress are shown again in Figure 7.17.

A typical case encountered for pipes in service is an elliptical or thumb-nail shaped defect located at the inner surface. For example, the fracture surface in chapter 3, Figure 3.6 is from a field failure of this type. Wang and Bell (2004) provide solutions for elliptical surface cracks of the type shown schematically in Figure 7.18.

Figure 7.19 shows the corresponding *T*-stress solutions for elliptical defects of different crack lengths.

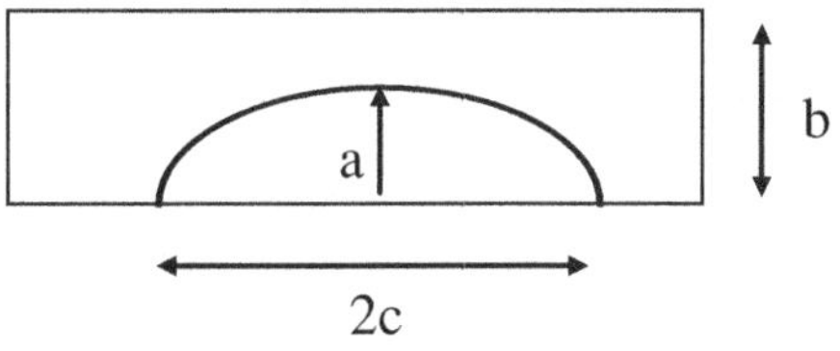

Figure 7.18 Elliptical surface crack (adapted from Wang and Bell, 2004)

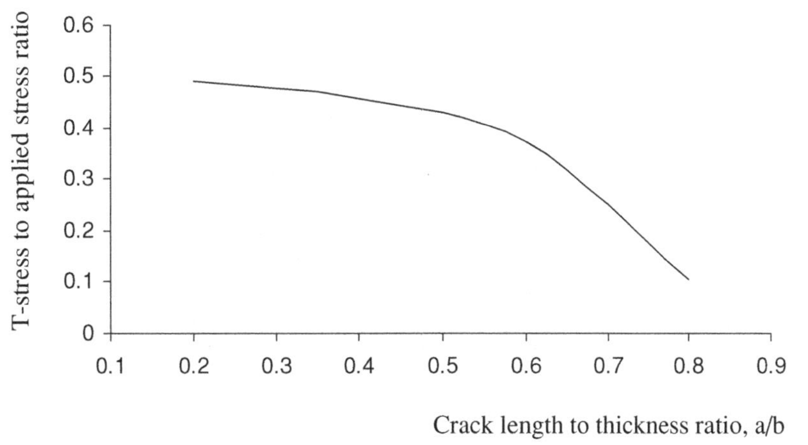

Figure 7.19 *T*-stress solutions for elliptical defects of different lengths. Aspect ratio a/2c = 0.4. (data from Wang and Bell, 2004)

Table 7.3
Remote applied stress and corresponding *T*-stresses for hypothetical example

Material	Remote applied stress	*T*-stress (assumed for $a/b < 0.2$)
PE80C (AUS)	1,160.5 psi (8.0 MPa)	580.3 psi (4.0 MPa)
PE80B (AUS)	1,160.5 psi (8.0 MPa)	580.3 psi (4.0 MPa)
PE100 (AUS)	1,450.6 psi (10.0 MPa)	725.3 psi (5.0 MPa)
PE3408 (US)	1,600 psi (11.03 MPa)	800.0 psi (5.5 MPa)

As an example, consider a set of PE80C (AUS), PE80B (AUS), PE100 (AUS) and PE3408 (US) pipes, each containing an elliptical crack with length to thickness ratio $a/b < 0.2$ at the inner surface. For each pipe in this hypothetical example, Figure 7.19 indicates that the ratio of *T*-stress to remote applied tensile stress (perpendicular to the crack plane) is approximately 0.5 (Wang and Bell, 2004). As a worst case, it is assumed that the circumferential tensile stress in the wall of each pipe is equal to that specified in the long term hydrostatic strength category (or hydrostatic design basis for PE3408 (US)). Referring back to Tables 1.2 and 1.3 in chapter 1, this corresponds to the remote applied stresses and corresponding *T*-stresses in Table 7.3.

Therefore, by extracting values of σ_Y and σ_D from Figures 7.13 to 7.16, the deformation state at the tip of the defect in each PE material can be determined using the *T*-stress values in Table 7.3. Figure 7.20 shows the different crack tip deformation states that would exist based on σ_Y and σ_D values at 1,000 hours.

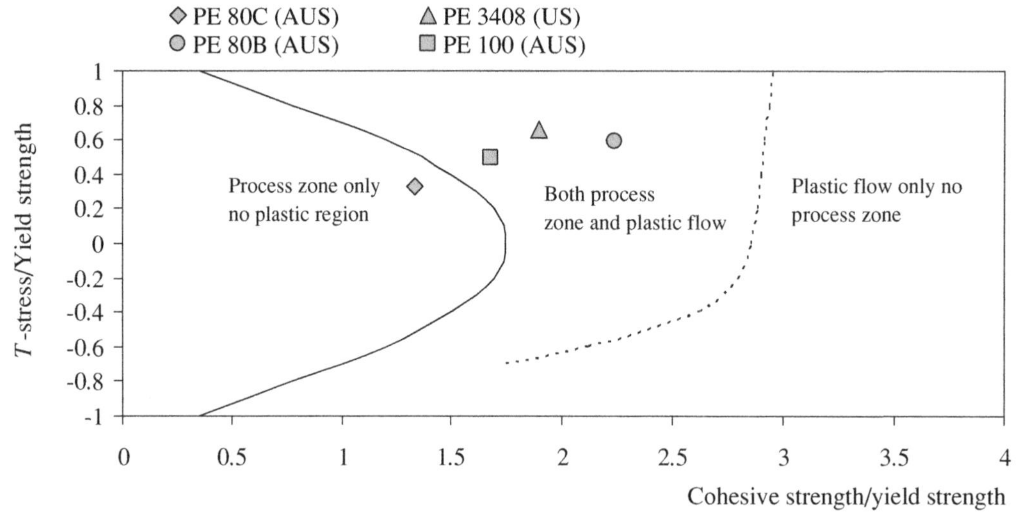

Figure 7.20 Long-term crack tip deformation states for different PE materials. Hypothetical case of a 4 inch DR 11 pipe containing an elliptical surface crack at the inner surface. Internal pressure = 87 psi. Crack tip deformation state criteria adapted from Broberg (1999).

For this particular example, Broberg's (1999) criteria for PE80B (AUS), PE100 (AUS) and PE3406 (US), suggest that the resulting crack tip craze will be embedded in a region of plastic flow. For these materials, LEFM theory is therefore invalid. Only PE80C (AUS) exhibits a crack tip process zone (or craze) in an elastic environment. In this particular case, Linear Elastic Fracture Mechanics (LEFM) theory may be valid, provided that the length of the crack tip craze does not exceed the maximum threshold level of 6% of the remaining ligament stipulated by Kanninen et al. (1990). For all materials, the time dependence of craze and yield strengths diverge further at extended times beyond 1000 hours. For a given *T*-stress, this means that the failure state positions in Figure 7.20 move to the right, which again invalidates the use of LEFM theory.

ELASTIC PLASTIC FRACTURE—MATERIAL *J-R* CURVES

As described in chapter 6, an analysis method for ductile fracture is the use of Elastic Plastic Fracture Mechanics (EPFM) theory. An engineering approach to predicting ductile fracture is based on the crack tip characterizing parameter known as the *J*-integral. As outlined by Williams (2000), J is an energy release per unit area of crack face for a non-linear elastic solid. However, the parameter J is also used for materials where a non-linear load-displacement relationship is brought about by plastic (rather than elastic) deformation. *J*-crack growth resistance values (often denoted as "J_R") can be determined following an experimental protocol described by the European Structural Integrity Society (Hale and Ramsteiner, 2000). The recommended test method uses either compact tension or SENB specimens, which are subject to a constant displacement rate until failure. As described in chapter 6, J_R values are determined from each test specimen by numerically integrating under the load vs. displacement plot resulting from each test. As outlined by Williams (2000), J_R is calculated as

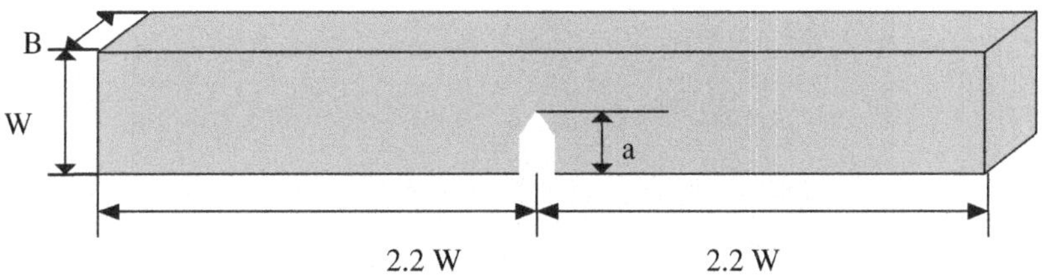

Figure 7.21 Single edge notched bend (SENB) specimen

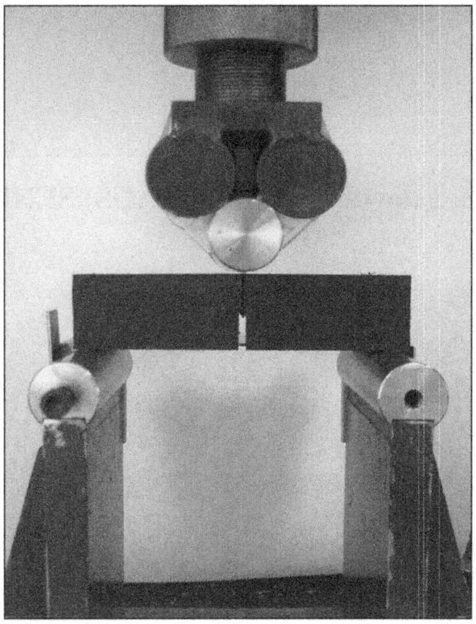

Figure 7.22 SENB test in progress

$$J_R = \eta \frac{U}{B(W - a_o)} \tag{7.3}$$

where U is the total energy determined by the area under the load-displacement plot; B is the specimen thickness; W is specimen width and a_o is the initial notch depth. η is a calibration factor, which for SENB specimens with $a_o/W = 0.5$, $\eta = 2$. To fully characterize J_R for a particular material, a series of specimens are loaded to different final displacements and the amount of crack extension that occurs during the test is measured. By plotting crack growth Δa against the corresponding J_R value, a material J-resistance (J-R) curve can be generated.

To determine the resistance to crack initiation, the Single edge notched bend (SENB) test was used. Following the ESIS protocol for the J-Integral test method, SENB specimens are machined to the dimensions shown in Figure 7.21. A point load is then applied directly opposite the notch while the specimen is simply supported at its ends. The photograph in Figure 7.22 shows an SENB test in progress.

Load-displacement curves for SENB specimens were determined using a screw-driven Instron testing machine in air at 20°C. These were then converted to material J-R curves (characterizing crack initiation resistance) following the ESIS protocol for J-integral testing (Hale and

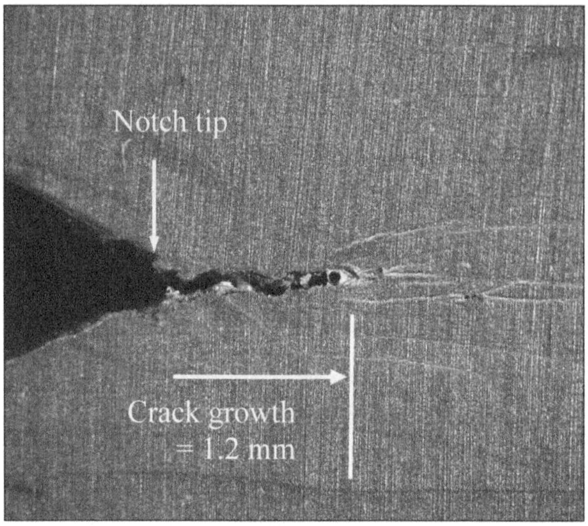

Figure 7.23 Example of crack growth in a PE80C (AUS) SENB sample

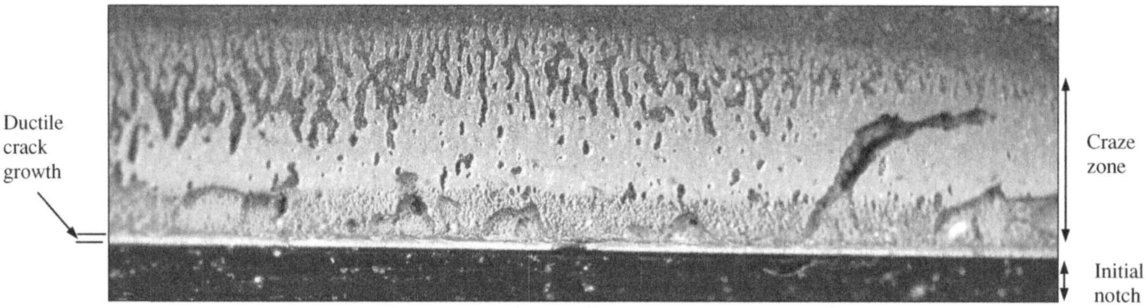

Figure 7.24 Fracture surface showing craze formation at the notch tip in an SENB test on PE80C (AUS)

Ramsteiner, 2000). To generate *J-R* curves for each PE material in this study, SENB specimens were machined with the following dimensions: Span, L = 160mm (6.3 inches); thickness, B = 20mm (0.8 inches); width, W = 40mm (1.6 inches); notch depth a_o = 20mm. The amount of crack growth generated in each test was measured by sectioning samples perpendicular to the direction of crack growth. As described in the ESIS protocol (Hale and Ramsteiner, 2000), this method allows the crack extension from the notch tip to be measured with reasonable accuracy. To validate these estimates, crack extension was also measured by completely fracturing specimens after exposure to liquid nitrogen (Bernal, Lopez, Montenegro, and Frontini, 2003). The photograph in Figure 7.23 shows a side view of crack growth in a PE80C (AUS) SENB sample.

Figure 7.24 shows the fracture surface obtained after a PE80C (AUS) sample is tested to failure in the Single Edge Notch (SENB) test.

As shown in Figure 7.24, the most dominant feature on the fracture surface is a large craze zone, approximately 2.0mm length, produced at the tip of the notch. This is indicated by the stress whitened zone, corresponding to the drawing of craze fibrils and local plastic deformation ahead of crack growth. A series of SENB tests on PE80C (AUS) all produced craze lengths of this size, which when compared to the un-notched ligament length of 20mm, exceeds the 6% threshold

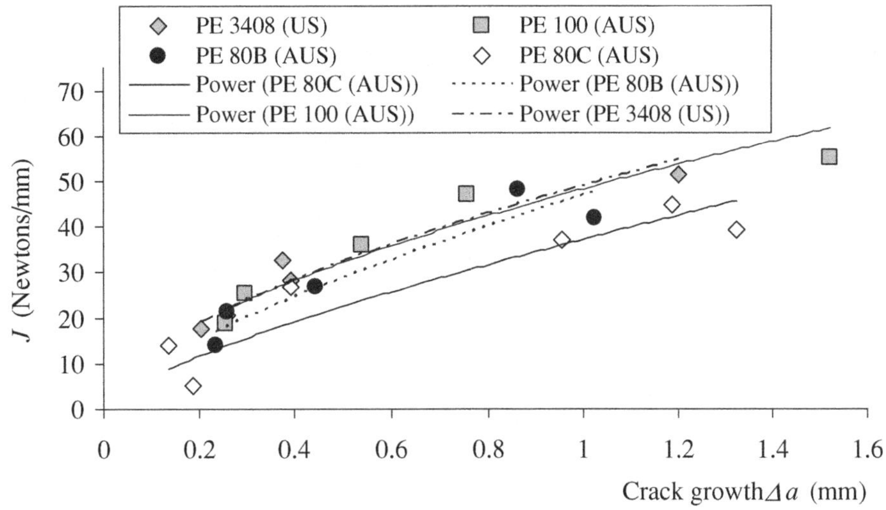

Figure 7.25 *J-R* curves for different PE materials in the SENB test

stipulated by Kanninen et al. (1990). This indicates that LEFM theory is also inapplicable for PE80C (AUS). Figure 7.25 compares measured *J-R* curves for each PE material examined.

As shown, PE3408 (US) and PE100 materials exhibit similar *J-R* curves, with higher applied *J*-values required to produce a given amount of crack growth. Although PE80B (AUS) shows a similar characteristic at low crack extensions, its *J-R* curve decreases at higher crack extensions. As with craze strength tests, the lowest resistance to ductile crack growth is exhibited by PE80C (AUS).

As described by Williams (2000), the resistance to ductile crack growth in tough polymers is characterised by fitting a power law in the form

$$J = A(\Delta a)^N \tag{7.4}$$

to the experimentally-determined *J-R* curves. The value of *J* at a crack extension of Δa = 0.2mm is chosen to characterise the resistance to crack initiation of a particular material. As reasoned by Williams (2000), although this crack extension is arbitrarily chosen, it is close to the lower limit of detectable values of Δa and is therefore a sensible definition of crack initiation. Table 7.4 compares power law parameters for material *J-R* curves and crack initiation *J*-values.

As shown in Table 7.4, crack initiation resistance $J_{0.2}$ values range from 11.5 N/mm (= 64.4 lbf/inch) for PE80C (AUS) to 19.0 N/mm (= 106.6 lbf/inch) for PE100 (AUS). $J_{0.2}$ values are in good agreement with values of 12.9 N/mm and 15.4 N/mm previously reported by Bernal, Lopez, Montenegro, and Frontini (2003) for pipe-grade PE materials.

Although the data in Figure 7.25 can be reasonably well represented using a power law, $J = A(\Delta a)^N$, the observed crack extension may simply be apparent and could be attributed to crack tip blunting caused by local yielding rather than true ductile crack growth. As described by Williams (2000), the theoretical blunting line is given by

$$J = 2\sigma_y \Delta a \tag{7.5}$$

Table 7.4
Material *J-R* curves and $J_{0.2}$ initiation resistance for different PE materials

Material	*J-R* curve *A* parameter	*J-R* curve *N* exponent	Crack initiation resistance $J = J_{0.2}$ (Newtons/mm) (lbf/inch)
PE80C (AUS)	37.3	0.73	11.5 (64.4)
PE80B (AUS)	42.5	0.63	15.4 (86.2)
PE100 (AUS)	48.4	0.58	19.0 (106.6)
PE3408 (US)	49.0	0.60	18.7 (104.5)

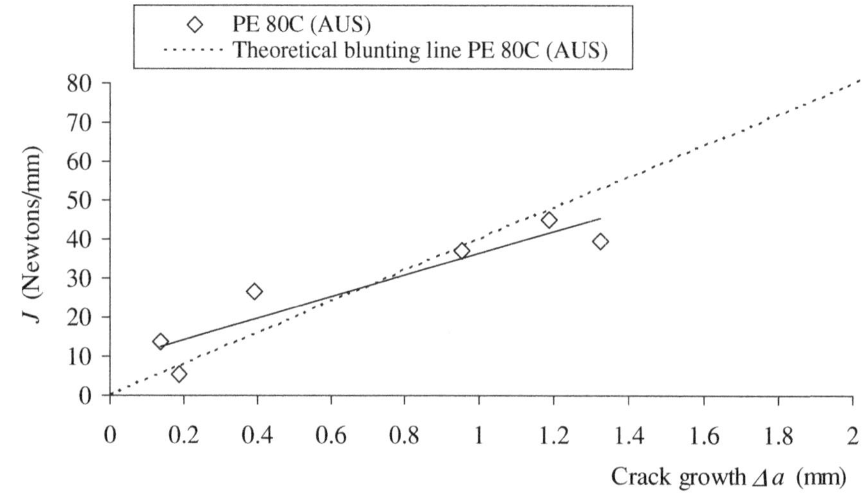

Figure 7.26 Comparison of *J-R* data with theoretical blunting line for PE80C (AUS)

where σ_Y is the material yield strength. Figures 7.26 to 7.29 show comparisons between measured J vs. Δa data and the theoretical blunting line, which accounts for the apparent increase in crack length due to crack tip blunting. A regression fit to each set of data points is also shown instead of the power-law fit.

As shown, despite the scatter of the data, the experimental points of each PE material lie reasonably close to their corresponding theoretical blunting lines. If actual crack growth had occurred, data would be expected to fall below the blunting line at high crack extensions (Hale and Ramsteiner, 2000). The absence of this departure may indicate that the apparent crack growth in these tests is actually due to local yielding and collapse of material at the crack tip, which has been previously reported to occur in tough PE grades (Bernal, Lopez, Montenegro, and Frontini, 2003).

CONCLUDING REMARKS

This chapter has identified and measured the relevant fracture properties of PE materials that influence long-term fracture failure in the field. It should be noted that the properties measured in this chapter apply to this particular set of samples only and should not be assumed as typical for *all* materials of that particular designation. However, as a first approximation, results from these tests can be used to compare the performance of different materials, and to provide

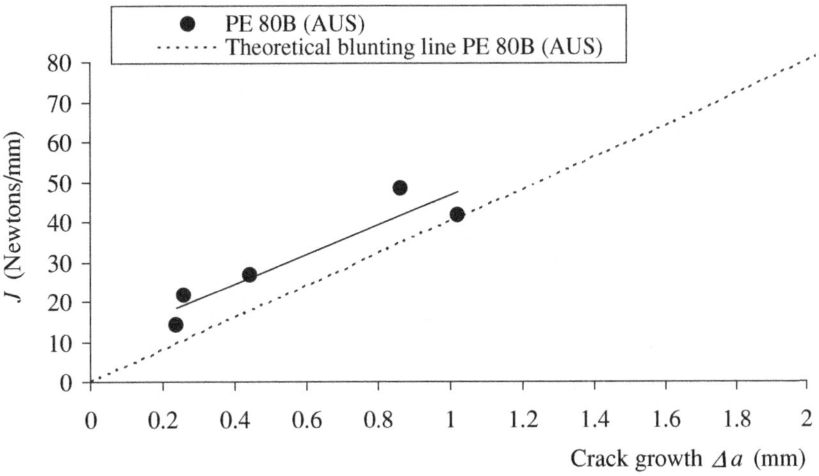

Figure 7.27 Comparison of *J-R* data with theoretical blunting line for PE80B (AUS)

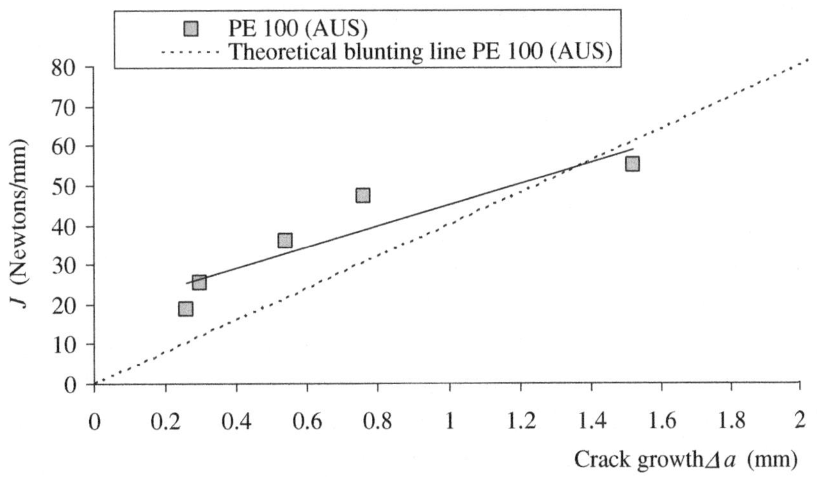

Figure 7.28 Comparison of *J-R* data with theoretical blunting line for PE100 (AUS)

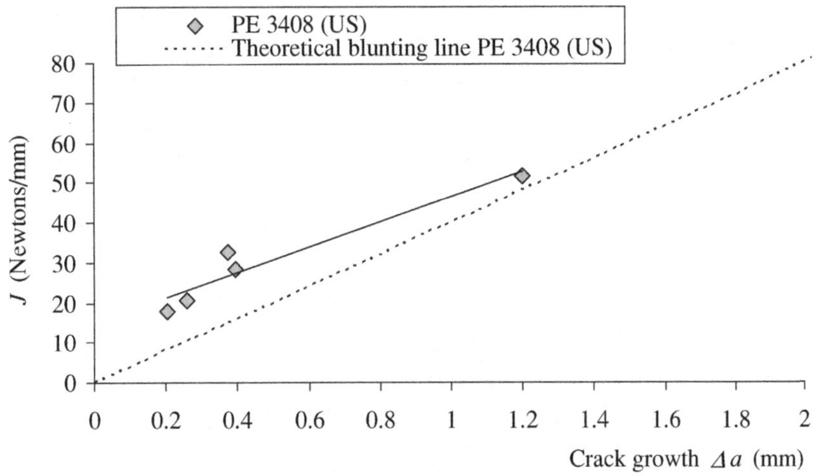

Figure 7.29 Comparison of *J-R* data with theoretical blunting line for PE3408 (US)

input to long-term fracture failure models. Details of materials tested are summarized in the following table:

Sample	Origin	PE grade	Relevant Standard	Hydrostatic strength category (or Hydrostatic Design Basis) at 50-year intercept
PE80C (AUS)	Australia	PE80C	AS 4131	8.0 MPa (1,160 psi)
PE80B (AUS)	Australia	PE80B	AS 4131	8.0 MPa (1,160 psi)
PE100 (AUS)	Australia	PE100	AS 4131	10.0 MPa (1,450 psi)
PE3408 (US)	USA	PE3408	AWWA C909	11.0 MPa (1,600 psi)

Circumferentially Deep Notched Tensile (CDNT) tests were conducted to isolate and examine the craze separation behaviour in each material. To assess the resistance to craze separation under constant applied stress conditions, CDNT tests were conducted using a lever-arm rig with different static loads. Time to failure was recorded in each case allowing the variation of craze stress with time to be quantified. The ranking of PE materials in terms of long-term craze resistance (highest to lowest) is:

1. PE100 (AUS)
2. PE3408 (US)
3. PE80B (AUS)
4. PE80C (AUS)

To check the validity of Linear Elastic Fracture Mechanics (LEFM) theory for each of these PE materials, static craze strength and uniaxial yield strength data were combined in Broberg's (1999) assessment criteria for the stress state ahead of a defect or a crack. Results indicate that for all materials, LEFM theory is invalid.

An alternative to LEFM is to use Elastic Plastic Fracture Mechanics (EPFM) theory to analyze ductile fracture. An engineering approach to predicting ductile fracture is based on the crack tip characterizing parameter known as the J-integral. J-R curves were generated by plotting values of J that were measured for different levels of crack growth. PE3408 (US) and PE100 (AUS) materials exhibit similar J-R curves, with the highest J-values required to produce a given amount of crack growth. As with craze strength tests, the lowest J values (and hence resistance to ductile crack growth) is exhibited by PE80C (AUS). Although the data could be reasonably well represented using a power law, $J = A(\Delta a)^N$, comparisons with theoretical blunting lines were also made. The theoretical blunting line is given described by the relationship $J = 2\sigma_y \Delta a$ and represents the *apparent* increase in crack length from blunting at the crack tip due local yielding and collapse. Despite the scatter of the data, the experimental points of each PE material lie reasonably close to their corresponding theoretical blunting lines. Agreement with theoretical blunting lines may indicate that a fully plastic regime exists at the crack tip in these materials. In this regime, single-parameter fracture mechanics (i.e., J-based) may also be invalid and could not be transferred to other structures, such as pipes containing defects. In the case of a fully plastic regime, the failure mechanism is one of plastic collapse and the effect of a crack is simply to reduce the net cross-section of the structure.

CHAPTER 8
DEVELOPING A FAILURE MODEL FOR POLYETHYLENE PIPES

INTRODUCTION

The previous chapters of this report identified and measured the relevant fracture properties of Polyethylene (PE) pipe materials. This chapter details how this information can be used to develop a fracture failure model for a PE pipe subjected to in-service loading conditions. Other failure types, such as those from poorly installed joints and fittings, are the result of poor installation practices and are not addressed in the model. As described below, three different failure models are proposed to capture the behavior of a PE pipe subjected to combined internal pressure and diametrical deflection. Predicted failure times from each model are compared to those obtained from an experimental test arrangement, allowing the most accurate failure model to be selected for further development in chapter 9.

ELASTIC-PLASTIC FRACTURE MECHANICS MODEL

The analysis conducted in chapter 7 established that Linear Elastic Fracture Mechanics (LEFM) theory cannot be applied to any of the PE materials assessed in this study. An alternative method to predict ductile fracture failure is based on Elastic Plastic Fracture Mechanics (EPFM) theory and addresses the balance between the applied crack driving force *J* and the material resistance to ductile fracture. However, the comparison of material *J-R* curves and theoretical crack blunting lines in chapter 7 suggested that actual ductile crack growth may *not* take place in the PE materials currently under investigation in this study. Therefore, a second check on the validity of EPFM theory to these materials can be conducted by applying predictions from an EPFM failure model to experimental observations of failure times on the laboratory. As described in chapter 6, for an internally pressurized cylinder with a radial crack at the inner surface, the corresponding elastic-plastic crack driving force (*J*) is expressed as (Kumar, German, and Shih, 1981).

$$J = \frac{K_{ITOT}^2}{E} + \alpha\varepsilon_o\sigma_Y c h_1\left\{\left(\frac{p}{p_0}\right)^{n+1}\right\} \tag{8.1}$$

K_{ITOT} is the total applied Stress Intensity Factor (SIF) for in-service loading conditions and can be calculated from LEFM theory; E is the short-term material tensile (or secant) modulus; c is the uncracked ligament length; p is the applied internal pressure; σ_Y is the material yield strength and the parameters α, ε_0 and n are obtained from the stress-strain relationship for the material. The dimensionless EPFM function h_1 is tabulated for a range of crack lengths and pipe sizes in the GE Handbook (Kumar, German, and Shih, 1981).

Stress-Strain Relationship for PE Materials

The GE handbook (Kumar, German, and Shih, 1981) contains *J* solutions for materials that exhibit material stress-strain relationships in the form

Table 8.1
Ramberg-Osgood stress/strain parameters for PE materials

	Short-term tensile modulus, E MPa (psi)	σ_0 (MPa and psi)	ε_0	λ	n
PE80C (AUS)	721.3 (104.6×10^3)	19.5	0.02	1.9	1.6
PE80B (AUS)	703.8 (102.1×10^3)	17.0	0.02	2.9	1.6
PE100 (AUS)	940.4 (136.4×10^3)	18.5	0.02	3.4	1.6
PE3408 (US)	1,032.1 (149.7×10^3)	17.7	0.02	4.7	2.0

$$\frac{\varepsilon}{\varepsilon_0} = \lambda\left(\frac{\sigma}{\sigma_0}\right)^n \tag{8.2}$$

where ε is the plastic strain and σ is stress. ε_0, σ_0 and λ are material constants that can be extracted from experimentally determined stress-strain curves. Equation 8.2 represents Ramberg-Osgood's law which is widely adopted in the description of stress-strain behaviour of metals. However, it has also been previously used to describe the behaviour of pipe-grade PE materials (Bernal, Lopez Montenegro, and Frontini 2003). Table 8.1 summarises the relevant stress-strain parameters and material properties that are required to implement the EPFM analysis in Equation 8.1.

It should be noted that the measured values in Table 8.1 strictly apply to the set of samples only. Industry expert opinion suggests that there may be wide variation within an individual strength category (Stahmer, 2006). However, to illustrate model development, the values in Table 8.1 are used.

Applied Stress Intensity Factor (SIF) for In-Service Operating Conditions

For a buried pipe in service, the total applied SIF (K_{ITOT}) can be split into components from internal pressure (K_{IP}), diametrical deflection loads (K_{ID}) and processed-in residual stress (K_{ID}). The total applied SIF is given by

$$K_{ITOT} = K_{IP} + K_{ID} + K_{IR} \tag{8.3}$$

For a pipe subjected to internal pressure only, the SIF associated with an internal defect is written as (Lu, Davis, and Burn, 2004):

$$\frac{K_{IP}}{\sigma_{IP}(\pi a)^{1/2}} = Y_P = 1.824 - 4.669\left(\frac{a}{b}\right) + 14.959\left(\frac{a}{b}\right)^2 - 6.341\left(\frac{a}{b}\right)^3 \tag{8.4}$$

where a is the radial depth of a defect from the pipe inner surface, and b is the pipe wall thickness. Both a and b are measured in metres. Following LEFM notation, Y_P is the geometric correction factor for this geometry and loading type. σ_{IP} is the hoop stress at the pipe inner surface, given by the Lamé equation for a thick-walled cylinder:

$$\sigma_{IP} = \frac{p(R_i^2 + R_o^2)}{(R_o^2 - R_i^2)} \tag{8.5}$$

Here, p is the internal pressure (in Pascals), and R_i and R_o are the pipe inner and outer radii (in meters) respectively. The applied SIF for deflection loading is (Lu, Davis, and Burn, 2004):

$$\frac{K_{ID}}{\sigma_{ID}(\pi a)^{1/2}} = Y_D = 0.803 + 0.353\left(\frac{a}{b}\right) - 1.640\left(\frac{a}{b}\right)^2 - 1.003\left(\frac{a}{b}\right)^3 \tag{8.6}$$

In this case, σ_{ID} is the remotely applied membrane hoop stress at the location of the defect at the pipe inner surface.

The applied SIF associated with processed in residual stress is (Lu, Davis, and Burn, 2004):

$$\frac{K_{IR}}{\sigma_{IR}(\pi a)^{1/2}} = Y_R = 0.727 + 2.791\left(\frac{a}{b}\right) - 2.214\left(\frac{a}{b}\right)^2 - 2.149\left(\frac{a}{b}\right)^3 \tag{8.7}$$

where σ_{IR} is the magnitude of residual stress at the pipe inner surface (in Pascals).

Crack Tip Process Zone

The applied SIF K_{ITOT} and the dimensionless factor h_I in Equation 8.1 both depend on crack length. To account for the crack tip craze zone, EPFM theory defines an "effective" crack length as the original crack length a plus an additional length ϕr_y, which is some fraction of the crack tip craze zone.

$$a_e = a + \phi r_y \tag{8.8}$$

As outlined by Bernal, Lopez Montenegro, and Frontini (2003), the additional crack length is given by

$$\phi = \frac{1}{1 + (p/p_0)^2}, \; r_y = \frac{1}{6\pi}\left[\frac{n-1}{n+1}\right]\left(\frac{K_{ITOT}}{\sigma_Y}\right)^2 \tag{8.9}$$

Although there is no EPFM-based expression to predict time–dependent crack growth directly, the effect of time can be incorporated indirectly through the change in craze zone length. As outlined by Atkins and Mai (1988), the time dependence of material yield strength σ_Y follows a power law in the form

$$\sigma_Y = \sigma_0 t^{-q} \tag{8.10}$$

As time progresses, σ_Y decreases as per Equation 8.10 and consequently, the length of the craze zone will increase following Equation 8.9. According to Equation 8.8, this will also produce an increase in the effective crack length a_e. Finally, through Equations 8.2 and 8.7, this increase in effective crack length will produce corresponding increases in the applied SIF K_{ITOT}. As outlined in the GE handbook (Kumar, German, and Shih, 1981), the dimensionless EPFM function h_I will also increase. Finally, referring back to Equation 8.1, the net effect of the time dependent decrease in yield strength is to increase the crack driving force J.

EPFM Criteria for Failure

In the elastic-plastic regime, crack initiation is assumed to occur when the crack driving force per unit width J exceeds the resistance to crack initiation J_C. After crack initiation, failure can occur by two possible mechanisms. Firstly, stable crack growth through the pipe wall can occur when the effective crack length calculated from Equation 8.8 exceeds the pipe wall thickness, b. Secondly, unstable crack growth can occur after some amount of stable crack growth in ductile materials. After crack initiation, unstable crack growth is predicted to occur at the point of tangency between the applied J vs. a curve (calculated from Equation 8.1) and the material J_R vs. Δa curve. At this point, the gradient of the two curves is equal and the amount of stable crack growth before instability can be determined. As outlined in chapter 6, the criterion for crack initiation and subsequent failure in the elastic-plastic regime are written as

$$\text{Crack initiation: } J = J_{0.2} = J_C \tag{8.11}$$

$$\text{Stable crack growth through pipe wall: } a_e = b \tag{8.12}$$

$$\text{Crack instability: } \begin{cases} J = J_R \\ \dfrac{\partial J}{\partial a} = \dfrac{dJ_R}{da} \end{cases} \tag{8.13}$$

At any given time, the crack driving force can be calculated from Equation 8.1 and compared with the criterion in Equations 8.11 to 8.13. The amount of stable crack growth prior to failure can also be calculated by solving the partial differential equation in Equation 8.13. While numerical methods are available for this solution, an estimated graphical solution can be obtained by simply observing the tangent point between the applied J vs. a and material J_R vs. Δa curves.

CRAZE MECHANICS MODEL

As outlined in chapter 6, the craze strength data obtained using the CDNT test (chapter 7) can be used to predict crack initiation using the concept of reference stress. According to Duan and Williams (1998), the reference stress is that stress which remains invariant during stress redistribution by creep in a particular structure. It was postulated that this value of the invariant stress serves as a reference stress for the whole structure and any uni-axial creep and rupture tests conducted at the reference stress could reproduce the behavior of the whole structure. An estimate of the upper bound of reference stress is

$$\sigma_{ref} = \frac{P\sigma_Y}{P_U} \tag{8.14}$$

where P is the current load on the component, P_U is the rigid-plastic collapse load for the component and σ_Y is the yield stress. It can be shown that for a pipe under internal pressure containing a radial crack at its inner surface, the reference stress is

$$\sigma_{ref\,(pressure)} = \frac{-p}{\ln([R_i + a]/R_o)} \tag{8.15}$$

where p is the actual applied internal pressure; R_i and R_o are the inner and outer pipe radii respectively and a is the radial crack length from the pipe inner surface. For the same pipe under diametrical deflection, it can be shown that the reference stress is

$$\sigma_{ref\,(deflect)} = \frac{\Delta E(b-a)}{0.584 D_m^2} \tag{8.16}$$

where Δ is the applied defection; E is tensile modulus; b is pipe wall thickness and D_m is the pipe mean diameter. Equation 8.16 describes rigid-plastic bending with a plastic hinge formed at the centre of the remaining un-cracked pipe wall. As a first approximation, it is assumed that the reference stress for a pipe under combined internal pressure and diametrical deflection is given by the sum of Equations 8.15 and 8.16.

$$\sigma_{ref\,(total)} = \sigma_{ref\,(pressure)} + \sigma_{ref\,(deflect)} \tag{8.17}$$

Duan and Williams (1998) propose that the craze stress measured under high constraint in the Circumferentially Deep Notched Tensile (CDNT) test can be equated to the reference stress at the tip of a crack in a particular component. Therefore, following Duan and Williams (1998), the theoretical relationship between crack initiation time and reference stress is based on experimental craze stress vs. failure time data as

$$t_f = \left(\frac{\sigma_{ref\,(total)}}{A}\right)^{-1/m} \tag{8.18}$$

where the parameters A and m describe time-dependent craze separation under static tensile stress and are given in Table 7.2 for different PE materials.

SIMPLE NET-SECTION COLLAPSE MODEL

The relatively close correspondence between the material *J-R* curves and theoretical blunting lines (Figures 7.26 to 7.29) may indicate that a fully plastic regime exists at the crack tip in the PE materials studied. When plastic collapse occurs before failure, a simple alternative to EPFM theory is to use limit load analysis. Limit load analysis assumes that the main effect of a crack is simply to reduce the net cross section of a structure. For a buried pipe under service loading conditions, failure is predicted to occur when the total applied stress exceeds the material yield strength. The net section yield stress, σ_{NS}, decreases linearly with increasing crack size according to:

$$\sigma_{NS} = \sigma_Y\left(1 - \frac{a}{b}\right) \tag{8.19}$$

where σ_Y is the time-dependent material yield strength according to the data in Figures 7.13 to 7.16 in chapter 7. For a buried pipeline in the field, plastic collapse of the remaining un-cracked ligament is assumed to occur when a crack has grown sufficiently long that σ_{NS} has decreased below the total applied stress, σ_{TOT}, which is the sum remote stresses due to internal pressure (σ_{IP}), diametrical (σ_{ID}), and residual stress (σ_{IR}). As shown previously in Equation 8.6, σ_{ID} is described in terms of the diametrical deflection (Δ) imposed on a pipe. In practice, Δ is usually unknown, but can be determined from the magnitude of external loads due to soil backfill, and any live surface load applied. Diametrical deflection in a buried flexible pipe is based upon the Marston Spangler theory from Iowa University and is given by:

$$\Delta = \frac{0.1 \times 10^{-3} D_m(\gamma H + P_{surf})}{\frac{8EI}{D_m^3} + 0.061 E_{soil}} \tag{8.20}$$

where D_m is the mean pipe diameter (in meters), γ is the unit weight of soil backfill (in kN/m^3), H is soil cover depth (in meters), P_{surf} is the external surface loading (in kPa), E is the tensile modulus of the pipe material (in MPa), and E_{soil} is the soil modulus (in MPa). I is the second moment of area of the pipe wall for ring bending and is given by:

$$I = \frac{b^3}{12} \tag{8.21}$$

COMPARING EXPERIMENTAL AND PREDICTED FAILURE TIMES

To validate the fracture model, a series of combined pressure/deflection tests were conducted on 110mm (4.3 inch) outer diameter SDR 11 PE80B pipe. A 775-mm (30.5-inch) long section of pipe with an axial notch of 150mm was sealed at each end with external end caps and supported by a central plate, as shown. The pipe cross-section was deflected by a rigid plate, which acted over a length of 400mm.

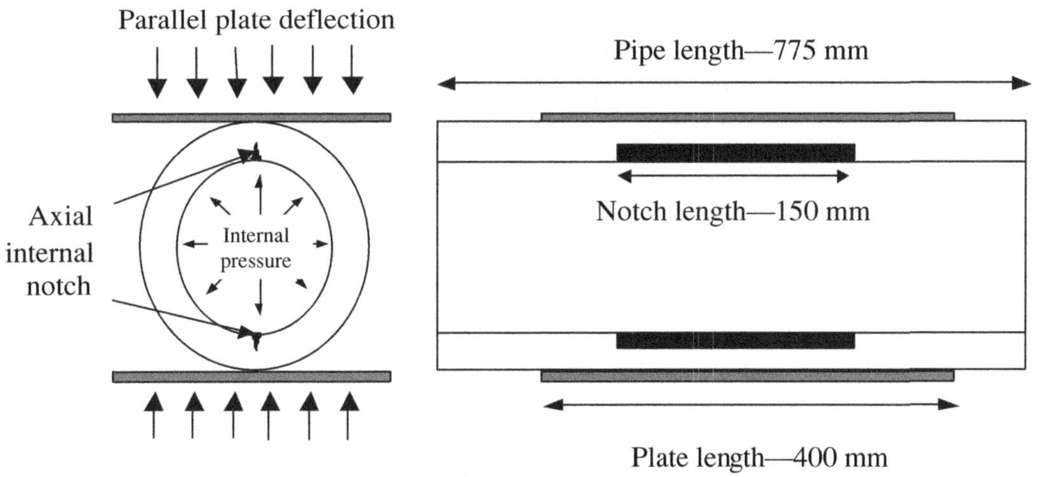

Figure 8.1 Experimental pressure and deflection loading arrangement

Table 8.2
Conditions for combined pressure/deflection testing of 110mm SDR 11 PE80B (AUS) pipe

Test	Material	OD (mm)	b (mm)	p (MPa)	Deflection Δ (mm)	Initial crack a_o (mm)	Time to failure t_f (hours)	Crack growth Δa (mm)
1	PE80B (AUS)	110.64	10.29	3.54	0.30	2.60	0.1	0.0
2	PE80B (AUS)	110.74	10.26	2.40	1.00	2.84	7.2	0.0
3	PE80B (AUS)	110.71	10.20	2.10	1.20	1.90	173.2	0.0

The schematic diagram in Figure 8.1 shows the experimental arrangement and test conditions and results are given in Table 8.2.

A comparison between observed failure times and predicted failure times from EPFM, craze mechanics and net-section yielding models is presented in Table 8.3.

As shown, the predicted failure times from the EPFM-based model consistently over-predict the time to failure with the largest error. This suggests that, (within the limits of experimental error on measured material properties) the failure mode in the actual tests was not one of ductile fracture. The invalidity of the EPFM based model is also suggested by the previous results in Figures 7.26 to 7.29 (chapter 7). Since SENB test data was well-represented by theoretical blunting lines it is assumed that crack extension is merely apparent in each PE material and the use of a single crack tip parameter analysis (i.e. J-based) is invalid.

In contrast to the EPFM-based model, predicted failure times from the net section collapse assumption are in better agreement. For the first test, the lowest error of –0.098 hours is achieved. As shown in Figures 8.2 and 8.3, the sample in test #1 did not fail by fracture but rather by yielding and drawing of material from the remaining ligament ahead of the notch tip.

The layer of drawn material shown is approximately 5mm wide and is comprised of yielded material that has been drawn from the un-notched ligament. Failure occurred when the drawn ligament ruptured, allowing pressurized water to escape.

Table 8.3
Comparing observed and predicted failure times for combined pressure/deflection testing of 110mm SDR 11 PE80B (AUS) pipe

Test	Actual fail time (hours)	EPFM Fail time (hours)	Err (hours)	Craze fail time (hours)	Err (hours)	Net section collapse time (hours)	Err (hours)
1	0.1	0.38	+0.28	0.02	–0.08	0.002	–0.098
2	7.2	1,100	+1,092.8	3.0	–4.2	0.1	–7.1
3	173.2	6,400	+6,226.8	110.1	–64.2	1.08	–172.8

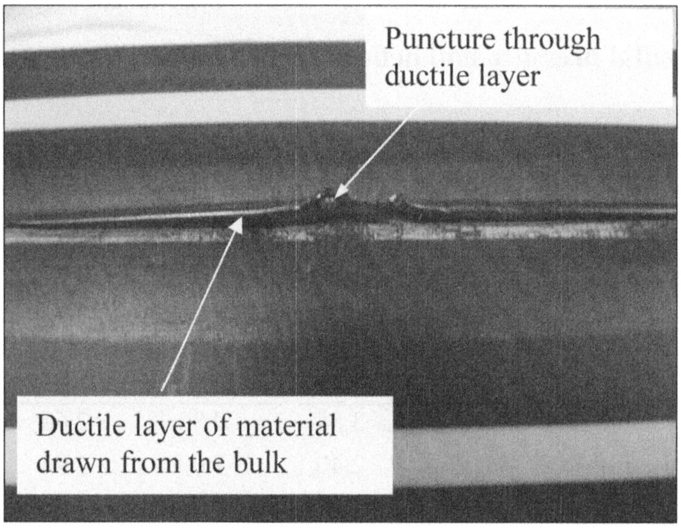

Figure 8.2 Ductile drawing failure mode observed in the combined pressure deflection test on pipe grade PE materials (image shows outer surface of PE80B (AUS) from test #2)

For the longer failure times in tests #2 and #3, the best agreement is achieved by the craze mechanics model (–4.2 and –64.2 hours for tests #2 and #3 respectively). Examination of failed samples from these tests show that although final failure occurred by yielding, drawing and rupture of material ahead of the crack tip, there is also evidence of crazing at the tip of the initial notch. This may explain why the craze failure model provides the best agreement in these cases. It should be noted however, that the craze failure model predicts the time to crack initiation from the initial defect and does not account for any subsequent crack growth that may occur. An alternative treatment would be model stepwise crack growth by formation and breakdown of successive crack tip crazes. For example, craze length could be calculated from crack length using Equations 8.8 and 8.9. The time to failure for the current craze length could then be predicted based from Equations 8.15 to 8.18. After this time, the crack is assumed to "jump" through the recently failed craze zone and grows in length accordingly. The new craze length can then be calculated and its failure time predicted. In this manner, stepwise crack growth through the pipe wall can be accounted for with the total failure time being the sum of individual craze failure times. However, initial calculations appear to significantly over-predict failure time and the craze initiation model still provides the most accurate, yet conservative prediction of failure time.

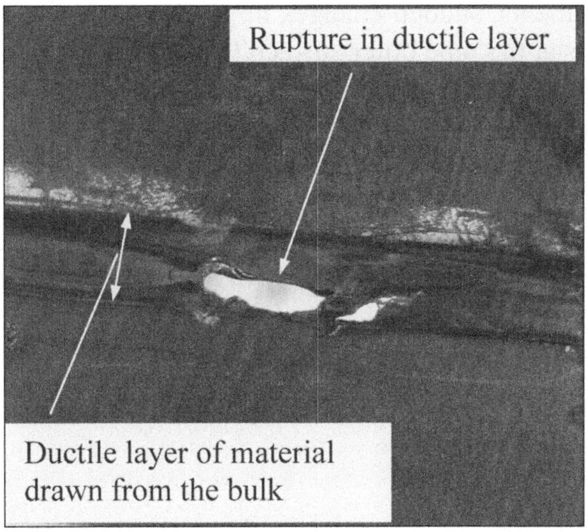

Figure 8.3 Ductile yielding and drawing of pipe wall in combined pressure deflection test on pipe grade PE materials (image shows outer surface of PE80B (AUS) from test #2)

CONCLUDING REMARKS

For the purpose of improving understanding of the long-term strength behavior of PE pipes important insights can be gained by conducting fracture mechanics evaluations on simplified models. In this chapter, a simplified fracture failure model has been developed to predict failure times for PE pipes under combined internal pressure and deflection loads. The method combines the time-dependent reduction in craze strength and the concept of an applied reference stress to predict an incubation period before crack initiation. Model predictions agree well with actual failure times for PE80B (AUS) over a range of internal pressures and diametrical deflection loads.

While the agreement between predicted and actual failure times is encouraging, further testing should be conducted to identify if a change in failure mode from ductile to brittle fracture occurs at extended test times. As discussed previously in chapter 6, Plummer, Goldberg, and Ghanem (2001) report a change in craze fibril texture from coarse (indicating ductile behavior) at high applied stress levels, to fine (indicating more brittle behavior) at lower applied stresses. However, it should be noted that Plummer, Goldberg, and Ghanem (2001) did not report transitions in more recent third generation pipe-grade PE materials such as those investigated in this study. Referring back to craze strength data in chapter 7, examination of Figures 7.13 to 7.16 indicates that for each PE material, long-term craze strength data can be represented by a single line. This suggests that no transition in the mode of craze separation (i.e., ductile to brittle) has occurred (Duan and Williams, 1998).

It should be noted that this model is merely a first approximation to the actual in-service loading conditions that a pipe can experience. For example, point loading (from inadequate installation) can create localized bending stresses at the inner surface of a pipe. Substantial in-service loads can also be generated by differential soil settlement, superimposed loads, and by thermal expansion effects (Mruk, 2006). While for simplicity these loading conditions are currently not included, their effect could be accounted for with appropriate reference stress solutions. Industry expert opinion also suggest that substantial "external defects," such as caused by surface scratches and gouges, rock impingements, sudden changes in geometry (as when a pipe enters a coupling) can lead to localized stress at the pipe outer surface (Mruk, 2006). However, since the processed-in residual stress at the outer surface of the pipe is compressive, these defects may be less detrimental than any that may be present within the pipe wall. For example, Sehanobish et al. (1985) and Chudnovsky, Sehanobish, and Wu (2000) report the presence of defects *within* the pipe wall at crack initiation sites in PE pipes.

It should also be noted that the model predicts the time to crack initiation from a defect and ignores the subsequent stepwise crack growth through the pipe wall. However, a more rigorous treatment is possible but initial calculations significantly over-predict failure times. Therefore, the craze initiation model provides the most accurate yet conservative prediction of failure time and is adopted for further development in chapter 9.

CHAPTER 9
ESTIMATING NETWORK-WIDE FAILURE RATES

INTRODUCTION

While modeling of individual pipe assets can provide information on the performance of those assets, water utilities often require performance predictions on a network-wide scale for asset management purposes. In theory, the characteristics for all pipe assets in a network could be input into the failure models developed in the previous chapter to predict when and where pipe failure would occur. In practice however, values for each of the relevant factors in the equations, such as external load, damage function, etc would need to be defined for each asset in the network. Obviously, much of this data is not available and valid estimations of these values based on uncertainty need to be made.

In order to extrapolate failure models from individual pipes to pipe networks, uncertainty in key variables must be included. The first step in such a probabilistic failure analysis is to formulate a simple "limit-state" failure criterion in terms of the load (L) and resistance (R) variables

$$m = R - L \tag{9.1}$$

where failure is defined to occur when $m < 0$. The limit state equation for crack extension through the pipe wall is given by

$$m = b - a_e \tag{9.2}$$

where b is the pipe wall thickness and a_e is the effective crack length given by Equations 8.8 and 8.10. Using craze mechanics and the reference stress concept, the criterion for crack initiation is given by

$$m = \sigma_{ref\,(total)} - \sigma_C \tag{9.3}$$

where $\sigma_{ref\,(total)}$ is the reference stress at the tip of a crack due to combined internal pressure and deflection and σ_C is the time dependent craze stress, given by Equation 7.1 with parameters in Table 7.2. Again, crack initiation is assumed to occur when $m < 0$.

As defined previously, a simple, alternative failure criterion for ductile fracture is the difference between the net section stress σ_{NS} and the total applied stress σ_{TOT} from operating loads:

$$m = \sigma_{NS} - \sigma_{TOT} \tag{9.4}$$

As described in chapter 8, σ_{NS} is given by Equation 8.19 and σ_{TOT} can be calculated using Equations 8.4, 8.6 and 8.19. Plastic collapse is assumed to occur when $m < 0$.

In all failure models that account for uncertainty, the basic variables that comprise the limit-state criteria (i.e., defect size, operating loads, etc.) are no longer single valued, but are treated as stochastic. This means their values follow probability distributions, which have mean values and variances associated with them. The limit-state failure criterion is reformulated in the form $M = f(\overline{Z})$, where M is the stochastic limit-state variable, and $\overline{Z}$ is a random vector of the stochastic variables that comprise the limit-state failure criteria (i.e., defect size, operating loads, etc.). Although probabilistic versions of these failure criteria (i.e., to estimate the probability that $M < 0$) have been developed, they often rely on simplifying assumptions (Ahammed and Melchers 1997). For example, Level II first-order second-moment (FOSM) methods are often used to account for uncertainty in limit-state models (Melchers 1999, Thoft-Christensen and Baker 1982). In FOSM methods, M is assumed to follow the normal probability distribution. Because of this assumption, there is a direct translation between the probability of failure (the probability that $M < 0$) and the shortest "distance" from the failure region defined in variable space and the origin (Melchers 1999). In reality, however, stochastic limit-state variables may not be normally distributed. Complex loading conditions also mean that limit-state failure models are rarely linear, and M is often a non-linear function of stochastic variables. This causes some degree of non-normality in M. Furthermore, iterative methods to calculate β require linear approximations to the failure surface, resulting in oversimplification. As a consequence of all assumptions, realistic physical failure mechanisms can be lost in order to obtain a solution.

A more straightforward way to estimate the lifetime probability functions (discrete probability density functions) is to use Monte Carlo simulation in conjunction with the limit-state formulations of pipe failure. Monte Carlo simulations sample lifetimes by repeatedly generating random numbers, and using these for calculating and evaluating the limit-state equations. For a simulated hypothetical network of pipes, a number of failures are recorded over time, which allows the lifetime distribution to be estimated. In contrast to FOSM methods, Monte Carlo simulations allow the details of the physical failure mechanism to be preserved.

A full description of the use of Monte Carlo simulation to estimate average failure rates in PVC pipes has been previously reported by Burn et al. (2005).

MODELING UNCERTAINTY IN STOCHASTIC VARIABLES

To implement a Monte Carlo simulation, stochastic variables that influence the limit-state failure criteria must be identified. For a buried pipe in service, possible stochastic variables are:

- Pipe size
- Fracture properties of pipe material
- In-service loading conditions
- Inherent defect size

It is reasonable to assume that in order to meet performance standards requirement, pipe size is well controlled during manufacture. Variation in pipe wall thickness and diameter can therefore be assumed to be insignificant. Although in-service loading conditions are uncertain, sufficient data to describe this uncertainty is currently available from water utilities. For plastic pipes, slow crack growth is most sensitive to changes in defect size (Barker, Bowman, and Bevis, 1983) and the inherent uncertainty in this variable must be quantified to implement a Monte Carlo simulation.

Probability Distribution of Inherent Defect Size in PE Pipes

The Weibull probability distribution has previously been shown to model the variation in inherent defect size for plastic pipes reasonably well (Burn et al. 2005). A Weibull variable is one which has a "survivor function," $S(a)$, defined as:

$$S(a) = \exp\{-(a/\alpha)^{\eta}\} \tag{9.5}$$

where $S(a)$ is the probability that the variable a (inherent defect size) is greater than or equal to a particular value, α is the scale parameter of the Weibull distribution, and η is the shape parameter. Following Crowder et al. (1991), an empirical survivor function can be obtained as outlined below.

Suppose that all measured inherent defects have been put in ascending order $a_{(1)} < a_{(2)} \ldots < a_{(n)}$, where n is the number of measured defects. The empirical $S(a)$ is given by:

$$\hat{S}(a)_i = 1 - \frac{i-1}{n} \tag{9.6}$$

where i is the position of the ith defect size in the list arranged in ascending order (Crowder et al. 1991). Taking the logarithms of Equation 9.5 indicates that if the Weibull distribution is appropriate, a plot of $\ln\{-\ln S(a)\}$ versus $\ln(a)$ should be roughly linear.

In a previous study, Burn et al. (2005) examined fracture surfaces from failed PVC pipes to identify and measure those inherent defects in the pipe wall that caused fracture failure. However, since fracture failures in PE do not occur frequently, very few fractured samples of PE pipe have been supplied from water utilities as part of the current study. In the absence of field failure samples, an alternative source of defect size data for PE pipes is available from a study conducted by Barker, Bowman, and Bevis (1983) on fatigue failure of PE pipes in the UK. As a first approximation, this data is used to develop a probability distribution for inherent defect size in PE pipes.

Figure 9.1 shows a Weibull plot for inherent defect size data for 110mm SDR 11 pipes taken from the study conducted by Barker, Bowman, and Bevis (1983). The data were obtained from SEM examinations of fracture surfaces from laboratory fatigue failures in UK MDPE pipes. As shown, a straight line can be used to represent the data with a regression coefficient of 0.97. This suggests that the Weibull distribution represents the variation in defect size reasonably well. From Equation 9.5, the gradient and intercepts of the straight line in Figure 9.1 allows the Weibull distribution parameters to be estimated. Values of η and α are 1.16 and 0.0004 respectively, which result in an expected maximum defect size of 0.0038 inches and a standard deviation of 0.003 inches.

The applicability of the Weibull Extreme value distribution to defect sizes in PE gas pipes has also previously been demonstrated by Chudnovsky, Baron, and Schulkin (1995). An examination of fracture surfaces from notched tensile tests indicated that defects located at crack initiation sites were variable in size and that this variation could be described by a Weibull-type probability distribution. For comparison with the data from Barker, Bowman, and Bevis (1983), Chudnovsky, Baron, and Schulkin (1995) reported a higher expected maximum defect size of 0.62mm (0.024 inches) and a higher standard deviation of 0.198mm (0.008 inches). Unfortunately, raw

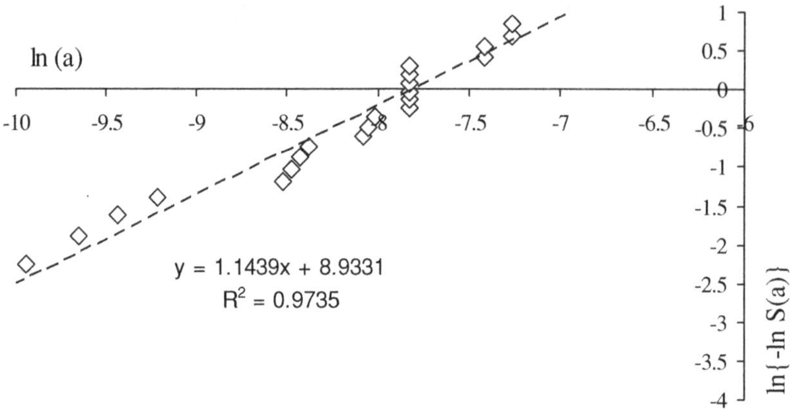

Figure 9.1 Weibull plot of inherent defect sizes for 110mm SDR 11 UK MDPE pipes (data from Barker, Bowman, and Bevis, 1983)

data required to calculate the corresponding Weibull scale and shape parameters is not available from this study (Chudnovsky, Baron, and Schulkin, 1995).

SIMULATING PE PIPE LIFETIMES USING MONTE CARLO SIMULATIONS

Having estimated the Weibull probability distribution for maximum inherent defect size *a*, Monte Carlo simulation can now be used to generate failures in a hypothetical PE pipeline. The steps below outline the computational procedure used in this study:

1. Create a population of short segments of PE pipes with known diameter and wall thickness.
2. Randomly assign an initial maximum defect size to each segment, based on the assumed Weibull probability distribution with scale and shape parameters $\alpha = 0.0004$ and $\eta = 1.16$ respectively.
3. Calculate remotely applied stresses for each pipe segment.

Time-marching loop

4. Increment time.
5. Calculate the limit-state equations for:
   a. Effective crack penetration though the pipe wall—Equation 9.2
   b. Failure based on craze mechanics—Equation 9.3
   c. Plastic collapse of un-cracked ligament—Equation 9.4
6. Record any pipe failures that occur, and replace failed segments with new segments.
7. Generate random inherent defect size in new pipe segments based on Weibull distribution as in Step 2.
8. Return to Step 4.
9. Write simulated lifetime data to output files after simulated time period ends.

The algorithm for Steps 1 through 9 has been implemented using the Python 2.3 programming language. A program listing, together with explanatory comments, is provided in appendix C.

Table 9.1
Input parameters for Monte Carlo simulation of 110mm SDR 11 PE pipe

Parameter	Value
Number of segments in pipeline	1000
Segment length (m)	1.0
Total simulated time (years)	500
Incremental time period (years)	1
EPFM crack initiation resistance, $J_{0.2}$ (N/mm)	From Table 7.4
Craze load bearing capacity A	From Table 7.2
Craze damage evolution parameter m	From Table 7.2
Inherent defect size—Weibull scale parameter α	1.16
Inherent defect size—Weibull shape parameter η	0.0004
Material short-term Young's modulus, E (MPa)	From Table 8.1
Ramberg-Osgood stress-strain parameter λ	From Table 8.1
Ramberg-Osgood strain parameter ε_0	From Table 8.1
Ramberg-Osgood stress parameter σ_0	From Table 8.1
Ramberg-Osgood stress-strain parameter n	From Table 8.1
Visco-elastic parameter for reduction in yield strength, q	0.04
Outer radius, r_0 (m)	From relevant pipe standard
Inner radius, r_i (m)	From relevant pipe standard
Maximum internal pressure in each segment, p (MPa)	See simulated cases
Soil cover depth, H (m)	See simulated cases
Soil unit weight, γ (kN/m^3)	19.0*
Soil modulus, E_{soil} (MPa)	6.0*
Surface load, P_{surf} (kPa)	See simulated cases
Residual hoop stress acting at the pipe inner surface, σ_{IR} (MPa)	3.1†

* Estimate based on un-compacted sandy soil (Standards Australia/Standards New Zealand, 1998).
† From Greenshields, Davis, and Leevers (1997), but it should be noted that residual stress will vary for different material grades (Stahmer, 2006).

RESULTS FROM MONTE CARLO SIMULATIONS OF PE PIPELINES

Using the Python program "PELifetime.py" (see appendix C), failure times have been generated for a series of hypothetical pipelines. Failure in each pipeline segment has been predicted using the models described by the limit state Equations 9.1 to 9.4 and the failure mode recorded. Relevant fracture properties for each PE material simulated have been assigned based on the experimental study outlined in chapter 7.

Simulated Failure Rates for PE Pipelines

Table 9.1 lists the input parameters used in the Monte Carlo simulation for PE pipelines.

A useful technique to derive failure rate versus age curves is to determine the probability distribution that best models the variation in simulated lifetimes. For example, Figure 9.2 shows

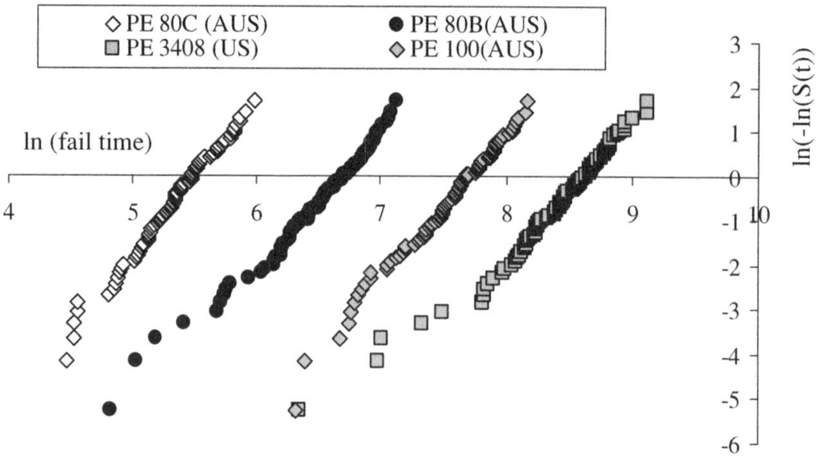

Figure 9.2 Weibull plot of simulated failure times in different PE materials. 110mm SDR 11 (AUS) PE pipes and DN 4 inch SDR 11 (US) PE pipes simulated.

Table 9.2
Weibull scale and shape parameters for simulated failure times in PE materials $p = 0.8$ MPa, $P_{surf} = 30$ kPa and $H = 0.5$ meters. All other input variables as in Table 9.1. 110mm SDR 11 (AUS) PE pipes and DN 4 inch SDR 11 (US) PE pipes simulated.

Material	Weibull scale parameter for failure time (α)	Weibull shape parameter for failure time (η)	Regression coefficient for Weibull fit (R^2)
PE80C (AUS)	248.6	3.2	0.96
PE80B (AUS)	814.7	2.9	0.98
PE100 (AUS)	5,603.5	2.8	0.97
PE3408 (US)	2,243.1	3.3	0.99

Weibull plots for simulated failure times in different PE pipe materials. For each material, 100 pipe segments were simulated under an internal pressure of 0.8 MPa (116 psi), an external surface load of 30 kPa (4.35 psi) and a soil cover depth of 0.5 metres (3.3 ft).

$S(t)$ is the empirical survivor function for failure time, as defined previously in Equation 9.6. As shown, lifetimes from Monte Carlo simulations are reasonably well represented by a straight line, illustrating the applicability of the Weibull distribution. The corresponding Weibull scale and shape parameters for failure time are listed in Table 9.2.

By making the simplifying assumption that all pipe segments along a pipeline have the same probability of failure (i.e., assuming homogeneous variation), the expected number of failures along a longer pipeline can be described by the binomial probability function (Crowder et al. 1991). This allows the expected failure rate (per unit pipe length/time period) to be estimated. If each individual pipe segment sustains one failure, which is independent of other failures, the expected number of failures in the pipeline is:

$$E(N) = nh(t) \tag{9.7}$$

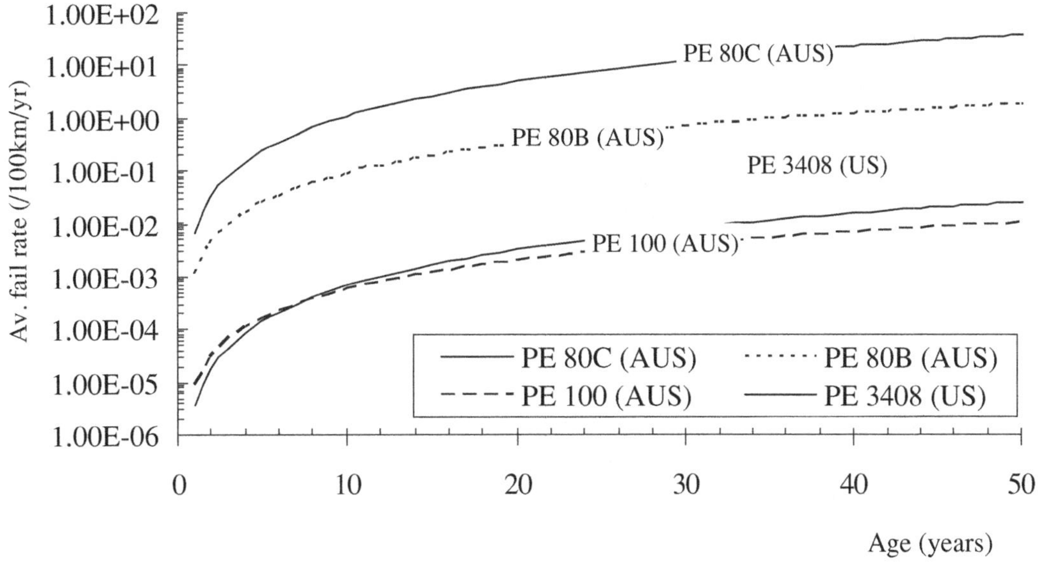

Figure 9.3 Average failure rates for 110mm SDR 11 (AUS) PE and DN 4 inch (US) PE pipes. p = 0.8 MPa, P_{surf} = 30 kPa and H = 0.5 metres. Input variables as in Table 9.1.

where n is the number of segments in the pipeline, and $h(t)$ is the hazard/failure rate function (Crowder et al. 1991). In practice, $h(t)$ is a measure of the probability of imminent failure at time t, given that failure has not yet occurred (Crowder et al. 1991). For the Weibull distribution, $h(t)$ is defined in terms of the scale parameter α, and the shape parameter η as:

$$h(t) = \eta\alpha^{-\eta}t^{\eta-1} \tag{9.8}$$

Figure 9.3 shows the corresponding average failure rate vs. age curves for different PE materials using the Weibull scale and shape parameters for failure time in Table 9.2. Since the original defect data reported by Barker, Bowman, and Bevis (1983) for PE pipes was based on examination of 1-metre long pipe sections, the number of segments (n) in a 100km-long pipeline is 100,000.

As shown in Figure 9.3, PE80C (AUS) exhibits the highest average failure rate under the specified operating conditions, rising to approximately 20 per 100km/per year for 50 year old pipes. This is followed by PE80B (AUS) (1 per 100km/per year at age 50); PE3408 (0.1 per 100km/per year at age 50) and finally PE100 (AUS) (0.005 per 100km/per year at age 50). Figure 9.4 compares average failure rates from Monte Carlo simulations in each PE material with observed failure rates in UK PE80 and PE100 pipes from the UKWIR national failure database (MacKellar and Pearson, 2003). Raw failure rates for UK PE80 and PE100 pipes were previously presented in chapter 2 (Figure 2.9). Failure rates are plotted on a logarithmic scale.

Failure rates in the UKWIR database are generally in good agreement with the simulated failure rates for PE80C (AUS) and PE80B (AUS) materials. Simulated failure rates in both PE3408 (US) and PE100 (AUS) are significantly lower than the UKWIR data. However, these comparisons should be interpreted carefully, since PE80 and PE100 pipe materials in the UK may exhibit different craze strength characteristics than the materials investigated in this study. Furthermore, it should be noted that the UKWIR data contains failures reported in pipes of

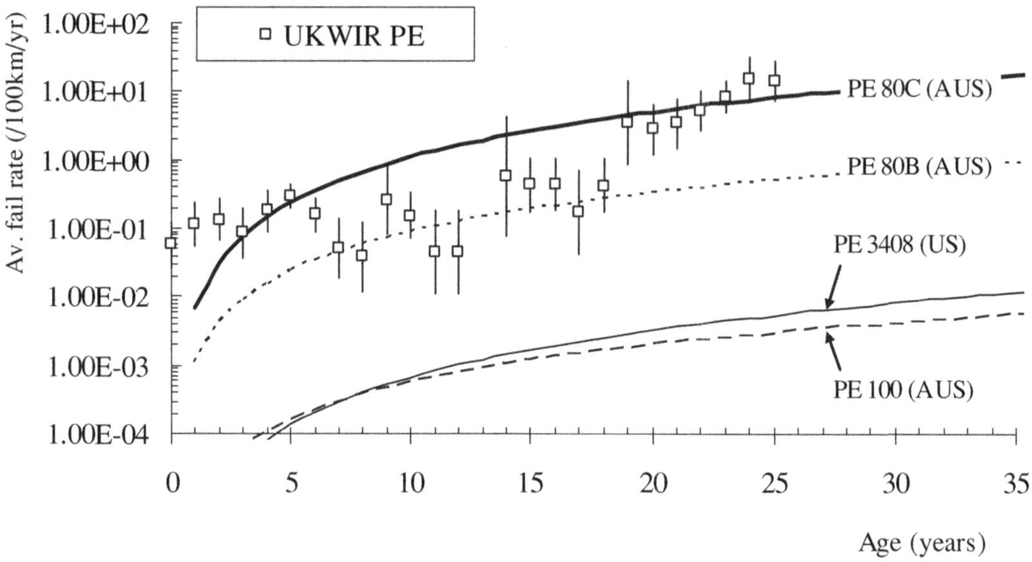

Figure 9.4 Simulated and average failure rates in the UKWIR database. p = 0.8 MPa, P_{surf} = 30 kPa and H = 0.5 metres. Input variables as in Table 9.1.

different diameters (between 50 and 500mm), operating at a range of unspecified internal pressures. Ideally, a more rigorous comparison should be conducted against recorded failure rates in PE80C/80B/100 (AUS) and PE3408 (US) 110mm SDR 11 pipes operating at known internal pressures.

The Effect of Pipe Operating Pressure

The calculations detailed above for the Weibull hazard function apply to an Australian 110mm SDR 11 (or a US DN 4 inch SDR 11) pipe operating at an internal pressure of 0.8 MPa. Table 9.3 gives Weibull scale and shape for different operating pressures.

Appendix D shows how the parameters in Table 9.3 can be used to calculated estimated values of expected failure rate in and example PE pipe. Figures 9.5 to 9.8 show the average failure rate vs. age curves corresponding to these parameters. Failure rates are plotted on a logarithmic scale.

As expected, the general trend is that simulated average failure rates decrease as pressure decreases. For example, in 100 year old PE80C (AUS) pipe, the failure rate decreases from 100 per 100km/per year at 0.8 MPa to 0.1 per 100km/per year at 0.5 MPa. However, historical failure data at different operating pressures is required to validate this trend. A similar effect of pressure is observed in the simulated average failure rates for other PE materials.

CONCLUDING REMARKS

To capture uncertainty in the pipe fracture process, a Monte Carlo simulation model has been developed for PE pipelines. Based on the craze separation and net section collapse models developed in chapter 7, the Monte Carlo simulation generates a hypothetical pipeline, comprised of a series of independent pipe segments. The maximum inherent defect size in each pipe segment is treated as a stochastic variable, which (based on raw data reported by Barker, Bowman, and Bevis (1983) is represented by the Weibull probability distribution function.

Table 9.3
Weibull scale and shape parameters for simulated failure times 110 SDR 11 (AUS) and DN 4 inch SDR 11 (US) PE pipe operating at different internal pressures

Material	Operating pressure in MPa and psi	Weibull scale (α)	Weibull shape (η)
PE80C (AUS)	0.5 (72.5)	1,021.8	3.9
	0.6 (87.0)	779.2	3.9
	0.7 (101.5)	563.5	3.7
	0.8 (116.0)	248.6	3.2
	0.9 (130.5)	110.8	3.2
PE80B (AUS)	0.5 (72.5)	2,594.8	3.9
	0.6 (87.0)	2,016.7	4.0
	0.7 (101.5)	1,455.7	3.7
	0.8 (116.0)	814.7	2.9
	0.9 (130.5)	279.6	3.2
PE100 (AUS)	0.5 (72.5)	8,423.0	3.9
	0.6 (87.0)	6,597.6	3.8
	0.7 (101.5)	5,027.1	2.7
	0.8 (116.0)	2,319.8	3.1
	0.9 (130.5)	1,076.3	2.7
PE3408 (US)	0.5 (72.5)	8,787.6	4.1
	0.6 (87.0)	6,765.1	4.0
	0.7 (101.5)	5,178.9	3.2
	0.8 (116.0)	2,243.1	3.3
	0.9 (130.5)	1,034.0	2.9

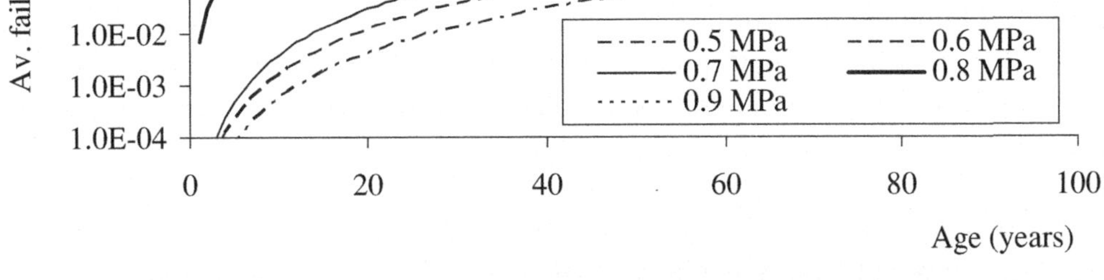

Figure 9.5 Simulated failure rates in 110 SDR 11 PE80C (AUS) pipes. For simulations, P_{surf} = 30 kPa and H = 0.5 metres. All other input variables as in Table 9.1.

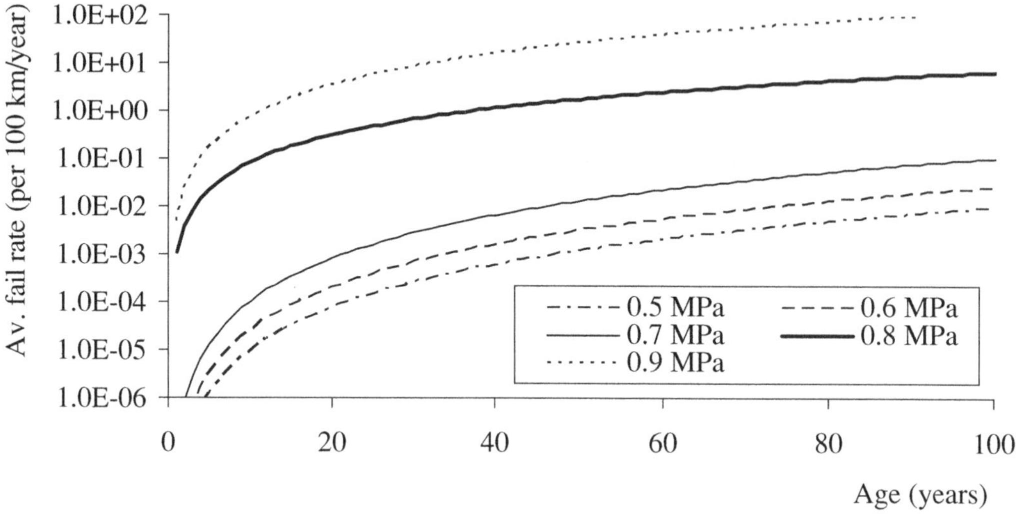

Figure 9.6 Simulated failure rates in 110 SDR 11 PE80B (AUS) pipes. For simulations, P_{surf} = 30 kPa and H = 0.5 metres. All other input variables as in Table 9.1.

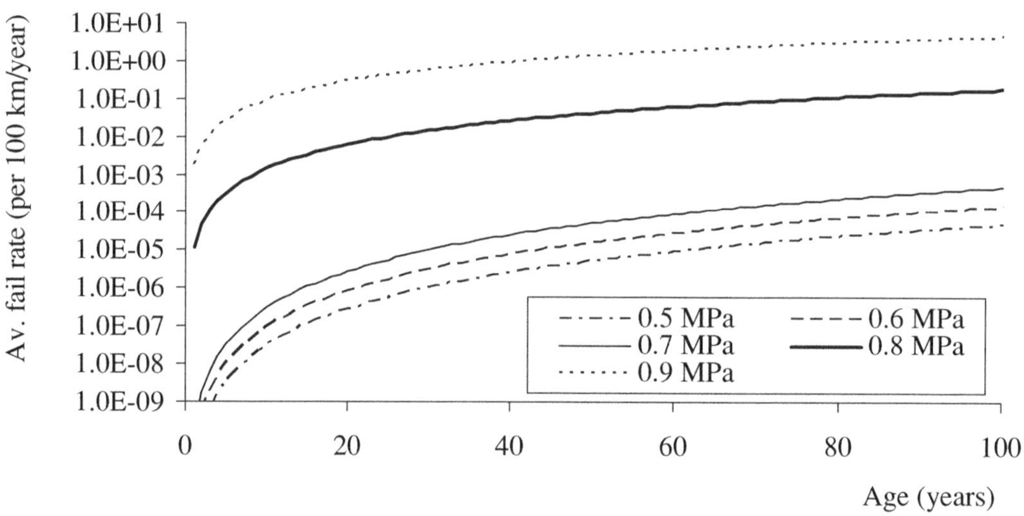

Figure 9.7 Simulated failure rates in 110 SDR 11 PE100 (AUS) pipes. For simulations, P_{surf} = 30 kPa and H = 0.5 metres. All other input variables as in Table 9.1.

For each segment in the hypothetical pipeline, the Monte Carlo simulation randomly assigns a maximum inherent defect size, based on the Weibull distribution. A time-marching loop is then performed in which the criterion for craze failure and plastic collapse is simulated in each pipe segment. Upon failure, a new pipe segment is installed and an inherent defect assigned. Making the simplifying assumption that the number of failures along the pipeline followed a binomial process, allowed the expected number of failures to be described by the Weibull hazard (or failure rate) function.

For comparison with actual failure rates recorded by water utilities, average failure rate vs. age curves were extracted from the UKWIR database. To ensure a fair comparison, only those failures that were reported as longitudinal fracture were extracted. Actual fracture failure rates in PE from the UKWIR database were compared to predicted failure rates for 110 SDR 11 PE pipes

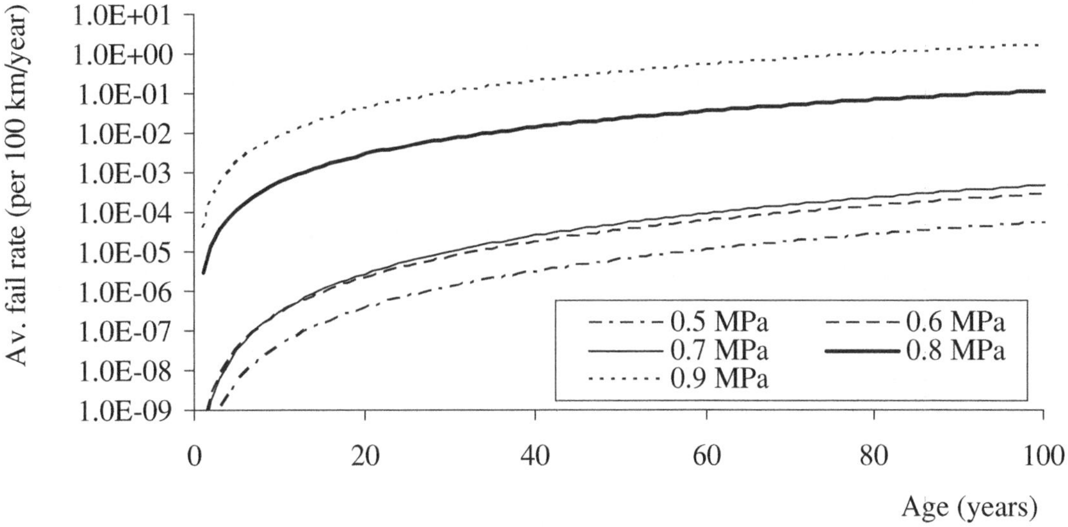

Figure 9.8 Simulated failure rates in DN 4 inch SDR 11 PE3408 (US) pipes. For simulations, P_{surf} = 30 kPa and H = 0.5 metres. All other input variables as in Table 9.1.

manufactured from PE80C (AUS) and PE80B (AUS) materials. While simulated curves showed reasonable agreement within the 95% confidence limits for actual failure rates, it should be noted that a relatively high operating pressure of 0.8 MPa (116 psi) was required to achieve this agreement. Furthermore, although the majority of historical failures in the UKWIR database were reported in pipes of diameters in the range 100 to 150mm, actual failure rates were was not restricted to 110 SDR 11 PE pipes. It should also be noted that actual operating pressure data is not included in the UKWIR database. Consequently, a more rigorous comparison of simulated and actual failure rates would require an extensive database of failures in 100 SDR 11 pipes with known operating pressures. Currently, such field performance data is not available for PE pipes in service. In comparison to PE80C (AUS) and PE80B (AUS) materials, PE100 (AUS) and PE3408 (US) exhibited significantly lower simulated average failure rates. This superior performance is attributed to the increased craze separation resistance and ductile fracture resistance exhibited by these PE materials. This leads to longer incubation periods before crack initiation and extended times before ultimate failure either by plastic collapse or ductile fracture. According to the simulations, even at relatively high operating pressure of 0.9 MPa (130.5 psi), an approximate failure rate of 1 per 100km/per year would be occur for pipes aged 100 years old. These forecast average failure rates in PE100 (AUS) and PE3408 (US) are extremely low compared to other water pipe materials used in the US and Australia (Rajani, McDonald, and Felio, 1995, Burn et al. 2005). However, it should be noted that until sufficient historical failure data becomes available these simulated failure rates cannot be validated.

Compared to other probabilistic modeling techniques, Monte Carlo simulation allows complex limit-state failure models to be used without the need for simplification. In contrast, techniques such as FOSM methods require linear approximations to failure models, and the assumption that stochastic variables are normally distributed. Consequently, the realism of the failure process can be compromised. However, while the Monte Carlo simulation model could be useful, several areas are identified for future development.

Inherent Defect Size

It should be noted that no actual defect size measurements were available for this study and data was taken from the literature for model development. Consequently, the applicability of the Weibull distribution for inherent defect size to PE pipes in Australia and the US is not yet validated. To this end, it is essential that pipe samples from USA/Canadian field failures should be exhumed and the fracture surfaces examined. While an initial examination under optical microscope will reveal the origin of crack initiation, SEM should be used to quantify the inherent defect size. It is also recommended that defect size databases are developed for different pipe diameters this will allow separate probability distributions to be developed for different pipe sizes.

Loading Conditions

The failure model requires the applied circumferential stresses due to internal pressure, external loading, and residual stress as input. While internal pressure and residual stress may be well defined, external loading is idealized in the model as a soil weight and a surface load acting above a buried pipe. A more realistic model should account for external loads generated from frost (in cold climates) and expansive clays (due to moisture depletion). For example, Rajani and Zhan (1996) propose an analytical model to estimate the magnitude of frost loads in cold climates. The model predicts the change in earth pressure as a frost front propagates, and shows good agreement with measured values. By considering the combined stiffness of a buried pipe and surrounding soil, it may be possible to calculate the magnitude of normal frost loads on a buried pipe, and bending stresses generated in the pipe wall. In a UK study, earth pressures in soil backfill adjacent to buried PVC-U pipes were recorded to increase by 15 kPa during winter. However, with minimum ground temperatures recorded at 6°C (43°F) (Marshall and Brogden, 1997), frost loading did not contribute to this load change. An alternative mechanism for this load increase may be the volumetric increase of expansive clay soils on moisture absorption. Again, depending on the surrounding soil support, this could create bending stresses in the wall of a buried pipe. A similar approach to that adopted for predicting frost loads may also be possible to model the action of expansive soils on buried pipes.

Since the Monte Carlo simulation method allows for a great deal of complexity in the failure models, more accurate descriptions of external loading under frost or expansive clays can be readily incorporated, provided analytical models can be developed.

Generic Failure Rate vs. Age Curves

The current study has demonstrated that the Weibull distribution function is well suited to describing the variation in lifetimes generated from the Monte Carlo simulation. Table 9.3 illustrates the pressure dependence of the Weibull scale and shape parameters that describe the lifetime distribution. However, it is reiterated that the model is strictly applicable to ***110 SDR 11 PE pipes only***, since this is the pipe size that the inherent defect size data applied (Barker, Bowman, and Bevis, 1983). Furthermore, it should be noted that the fracture properties used in the model apply strictly to the particular samples used in this study and ***should not be assumed to apply to all PE materials of that designation***. Further work is required to quantify the variation in material and fracture properties within different PE materials grades.

Although statistical techniques can be developed to extrapolate the defect size distribution to other pipe sizes, it is preferable to develop actual defect size data bases for different pipe sizes. As suggested above, these can be developed by exhuming and examining fracture failures in PE pipes of different diameters.

Extended Failure Times for PE100 (AUS) and PE3408 (US)

The relatively low failure rates in PE100 (AUS) and PE3408 (US) pipes are attributed to extended simulated failure times, which in turn, are attributed to the increased long term craze strength of these materials. However, other researchers suggest that the lifetime of a PE pipe under internal pressure should be divided into three stages (Andersson, 2001). During stage I, the pipe is mainly subjected to mechanical overload and ductile failure occurs by plastic instability. Stage II is often referred to as the "mechanical knee" in hoop stress vs. time data and corresponds to more brittle failures which occur from inherent defects in the pipe wall at lower applied stress levels. Stage III corresponds to the second knee in stress/time data and is referred to as the "chemical knee." This stage corresponds to the loss of the heat stabilizer package and the chemical life of the material under loaded conditions. Failures in stage III are exclusively brittle and failure time is independent of applied stress levels (Andersson, 2001).

The crack initiation/net section collapse failure model developed in this study does not account for chemical stabiliser loss and the onset of brittle failure. Therefore, identifying when stages II and III are activated at ambient temperature should reduce simulated failure times and increase average failure rates. In techniques such as the Rate Process Method (Palermo and DeBlieu, 1985), hydrostatic pressure testing at elevated temperatures is used to accelerate the onset of brittle failure in stages II and III. Empirical equations (usually in log form) are then used to estimate the transition to stages II and III in stress/time data at ambient temperature. However, as discussed in chapter 3, the microstructure of a PE pipe sample tested at 80°C differs from that of a pipe tested at ambient temperature and it is unclear whether the mechanism of crack growth under ambient temperature is preserved in tests at elevated temperatures. Further work should be conducted to identify the "chemical knee" and the onset of stage III brittle failures in these materials at ambient temperatures, should it exist.

APPENDIX A
PROGRAM LISTING FOR UKWIR PE FAILURERATES.PY

The calculation of average failure rates vs. age curves for UK PE pipes in chapter 6 has been implemented using the PYTHON 2.3 programming language. PYTHON is freeware, and can be installed from the Internet (www.python.org). The full listing for the program UKWIR PE FailureRates.py is given below. Text that is proceeded by "#" or "#*" is descriptive, and not part of the PYTHON code.

```
#***********************START OF PROGRAM******************************************

import os
import sys
import string
import math

_root = 'O://AWWARF PE project//PE Field performance//Python Scripts for failure rates'

def readPipeData(inpfname,dict):
   inpData  = file(inpfname)
   Query = inpData.readlines()
   inpData.close()
   for qrow in Query[1:]:
      if qrow != '':
         try:
            qrow = qrow.replace('"','')
            qrow = qrow.replace(' ','')
            try:
               var = string.split(qrow,',')
               id = int(var[0])
               installYr = int(var[4])
               length = float(var[7])
            except:
               pass
            dict[(id)] = [installYr,length]
         except:
            pass
      if qrow == '':
         pass
   return dict

pipeDict = { }
readPipeData(_root+'//PE failures (UKWIR database)(Mains)(Fracture failures Only).csv',pipeDict)

failureYears =
[0,1980,1981,1985,1987,1988,1990,1991,1992,1993,1994,1995,1996,1997,1998,1999,2000,2001,2002,2003]

#********************CALCULATE EXPOSURE***********************#

exposure = { }
for id in pipeDict.keys():
```

```
    iYear = pipeDict[id][0]
    length = pipeDict[id][1]
    for year in failureYears:
        age = year - iYear
        try:
            exposure[age] += length
        except:
            exposure[age] = length


def readBurstData(inpfname,dict):
    inpData  = file(inpfname)
    Query = inpData.readlines()
    inpData.close()
    for qrow in Query[1:]:
        if qrow != '':
            try:
                qrow = qrow.replace('"','')
                qrow = qrow.replace(' ','')
                try:
                    var = string.split(qrow,',')
                    count = int(var[10])
                    age = int(var[11])
                except:
                    pass
                dict[(count)] = age
            except:
                pass
        if qrow == '':
            pass
    return dict

burstDict = { }
readBurstData(_root+'//PE failures (UKWIR database)(Bursts)(Allpipes)(Fractures Only).csv',burstDict)

#*********************CALCULATE FAILURE AGES*********************#

failureTimes = { }
for id in burstDict.keys():
    t = int(burstDict[id])
    try:
        failureTimes[t] += 1
    except:
        failureTimes[t] = 1

#******************CALCULATE FAILURE RATES*********************#

failureRates = { }
for t in failureTimes.keys():
    try:
        failureRates[t] = float(failureTimes[t]) / float(exposure[t])
    except:
        pass
```

```
def wrtFailRate(fname,dict1,dict2):
   data = file(fname,'w')
   line = "%s" %(' ')
   line += "Pipe Age (years), Expected failure rate (per 100km/per year), # of observations, maxFailRate,
         minFailRate, +error, -error"
   for id in dict1.keys():
      line += "%s," %id
      failRate = float(dict1[id]*100e3)
      line += "%s," %failRate
      n = int(dict2[id])
      line += "%s," %n
      maxFR = failRate*math.exp(2.0/math.sqrt(n))
      minFR = failRate/math.exp(2.0/math.sqrt(n))
      line += "%s," %maxFR
      line += "%s," %minFR
      plusError = maxFR - failRate
      negError = failRate - minFR
      line += "%s," %plusError
      line += "%s" %negError
   data.write(line)
   data.close()


wrtFailRate(_root+'//PE failureRate vs Age(AllPipes)(Fractures Only).txt',failureRates,failureTimes)
```

APPENDIX B
POLYETHYLENE PRESSURE PIPES DESIGN FOR DYNAMIC STRESSES

INTRODUCTION

Polyethylene pressure pipes are designed on the basis of a burst regression line for pipes subjected to constant internal pressure. From this long term testing and analysis, nominal working pressure classes are allocated to pipes as a first indication of the duty for which they are suitable. However, there are many other factors which must be considered, including the effects of dynamic loading. Whilst most gravity pressure lines operate substantially under constant pressure, pumped lines frequently do not and it is essential that the effects of this type of loading be considered in the pipeline design phase to avoid premature failure. This note is intended to assist in the selection of pipe class for polyethylene pipes in applications involving transient and cyclic operating pressures.

Pressure fluctuations in pumped mains result from events such as pump start-up and shut-down or valves opening and closing. The approach adopted for pipe design and class selection when considering these events depends on the anticipated frequency of the pressure fluctuation as follows:

- For random, isolated surge events, for example, those which result from emergency shutdowns, the designer must ensure that the maximum and minimum pressures experienced by the system are within acceptable limits; and
- For frequent, repetitive pressure variations, the designer must consider the potential for fatigue and design accordingly.

DEFINITIONS

For the purposes of this note, surge is defined as a rapid, very short term pressure variation caused by an accidental, unplanned event such as an emergency shutdown resulting from a power failure. Surge events are characterized by high pressure rise rates with no time spent at the peak pressure. The maximum duration of a surge event is about 5 minutes.

Fatigue is associated with a large number of repetitive events. The key factors to consider are the size and frequency of the repeated event. For large pressure cycles, a lower number of events can be tolerated in the pipe lifetime. For smaller pressure changes, a greater number of events are acceptable.

The gradual diurnal pressure changes which occur in most distribution pipelines as a result of demand variation will not cause fatigue. The only design consideration required for this type of pressure fluctuation is that the maximum pressure should not exceed the pressure rating of the pipe.

SURGE DESIGN

It has long been recognized that polyethylene pipes are capable of handling short-term stresses far greater than the long-term loads upon which they are designed. However, this characteristic feature has, to date, not been utilized in design in Australia. Previous design recommendations have advised that the peak pressure should not exceed the nominal working pressure of the pipe.

Recently, studies in the UK (Marshall and Brogden, 1997) have examined the response of plastics pipes to short-term over-pressurization. This work has culminated in the publication of a design recommendation (Water UK Industry, 2000) allowing significant over-pressurization of plastics pipes during rapid surge events. However, while these results highlight the capabilities of plastics pipes in this respect, they do not discuss the effect of over-pressurization on the pipe where it is restrained by tapping bands or flange connections. The effect on other pipeline components is also not considered.

Field tests of installed pipelines are conducted to verify the pipeline design and construction. These tests are generally carried out at a pressure of 1.25 times the maximum system operating pressure. Since the test pressure is generally the highest pressure a pipeline will experience, the field test serves to demonstrate that all system components, including items such as anchor blocks which are formed in-situ, will be satisfactory for the operating conditions.

The demonstrated ability of polyethylene pipes to accommodate short-term surge pressures can be utilized to advantage by allowing some over-pressurization. However, it is recommended that the peak design pressure is limited to the field test pressure of that pipeline in order to ensure that the pipeline as a whole is capable of performing under these conditions.

Where the generation of negative pressures is anticipated, the possibility of transverse buckling should be considered. This topic is well covered in other technical literature.

FATIGUE DESIGN

Many materials will fail at a lower stress when subjected to cyclic or repetitive loads than when under static loads. This type of failure is known as fatigue. For thermoplastics pipe materials, fatigue is only relevant where a large number of stress cycles are anticipated. The important factors to consider are the magnitude of the stress fluctuation and the loading frequency. Where large stress fluctuations are predicted, fatigue design may be required where the total number of cycles in the operational lifetime of the pipe exceeds 100,000. For smaller stress cycles, a larger number of cycles can be tolerated.

The fatigue response of thermoplastics pipe materials has been extensively investigated (Joseph 1979, 1984; Strebel and Moet 1991; Beech et al. 1995; Bucknall and Dumpleton, 1985a, 1985b; Marshall, Brodgen, and Shepherd, 1998). The results of laboratory studies can be used to establish a relationship between stress range, defined here as the difference between the maximum and minimum stress (see Figure B.2), and the number of cycles to failure. From these relationships it is possible to derive load factors that can be applied to the operating pressures, to enable selection of an appropriate class of pipe.

This type of experimental data inevitably has a degree of scatter and it has been Australian practice, after Joseph (1984), to adopt the lower bound for design purposes. This approach is retained here because it ensures the design has a positive safety factor and recognizes that pipelines may sustain minor surface damage during installation, which could promote fatigue crack initiation. Note that for fatigue loading situations, the maximum pressure reached in the repetitive cycle should not exceed the static pressure rating of the pipe.

Table B.1 gives the recommended fatigue load factors for thermoplastics pipes. This is also graphically presented in Figure B.1.

Table B.1
Recommended fatigue load factors

Total cycles	Approx. no. cycles/day for 100 y life	PE80B and PE100 fatigue load factors	PE80C fatigue load factors
36,500	1	1.00*	1.00*
100,000	3	1.00	1.00
300,000	8	1.00	0.82
500,000	14	0.95	0.76
1,000,000	27	0.88	0.67
5,000,000	137	0.74	0.50
10,000,000	274	0.68	0.44
50,000,000	1,370	0.57	0.33

* Maximum allowable factor for fatigue loading pressure cycles.

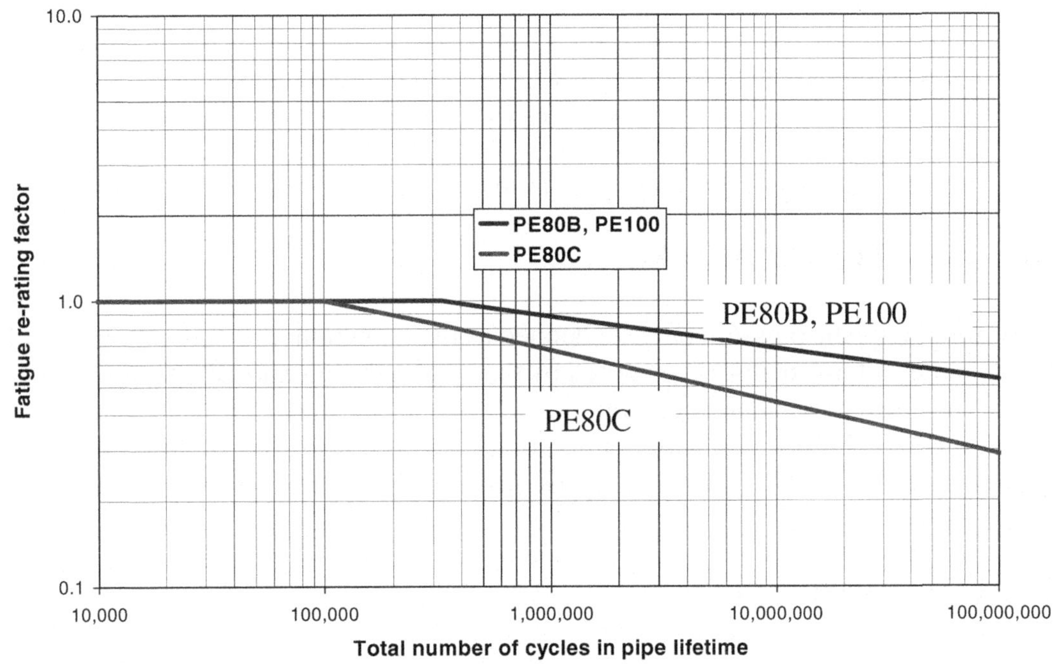

Figure B.1 Fatigue load factors for polyethylene pipes

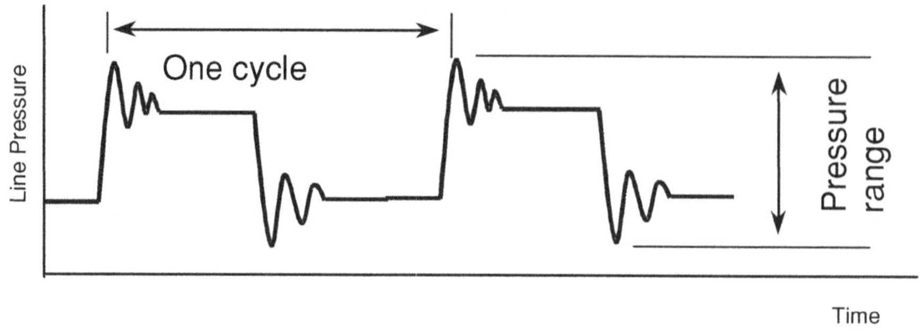

Figure B.2 Pressure cycle

PROCEDURE

To select the appropriate pipe class for fatigue loading, the following procedure should be adopted:

1. Estimate the likely pressure range, ΔP, i.e., the maximum pressure minus the minimum pressure.
2. Estimate the frequency or the number of cycles per day which are expected to occur.
3. Determine the required service life and calculate the total number of cycles which will occur in the pipe lifetime.
4. Using Table B.1 or Figure B.1, look up the fatigue load factor for the appropriate pipeline material and number of cycles.
5. Divide the pressure range by the fatigue load factor to obtain an equivalent operating pressure.
6. Use the equivalent operating pressure to determine the class of pipe required.

DEFINITION OF PRESSURE RANGE

For simplicity, the pressure range is defined as the maximum pressure minus the minimum pressure, including all transients, experienced by the system during normal operations as shown in Figure B.2. The effect of accidental conditions such as power failure may be excluded.

Figure B.2 also illustrates the definition of a cycle as a repetitive event. In some cases, the cycle pattern will be complex and it may be necessary to also consider the contribution of secondary cycles.

COMPLEX CYCLE PATTERNS

Pumping systems are frequently subject to surging following the primary pressure transient on switching. Such pressure surging decays exponentially, and in effect, the system is subjected to a number of minor pressure cycles of reducing magnitude. In order to take this into account, the effect of each minor cycle is related to the primary cycle in terms of the number of cycles which would produce the same crack growth as one primary cycle.

Using this technique, it is shown by Joseph (1984) that a typical exponentially decaying surge regime is equivalent to 2 primary cycles. Thus for design purposes, the primary cycle amplitude only is considered, with the frequency doubled.

In general, similar technique may be applied to any situation where smaller cycles exist in addition to the primary cycle. Empirically crack growth is related to stress range according to $(\Delta\sigma)^{3.2}$. Thus, n secondary cycles of magnitude $\Delta\sigma_1$ may be deemed equivalent in effect to one primary cycle, where

$$n = \left(\frac{\Delta\sigma_0}{\Delta\sigma_1}\right)^{3.2}$$

For example, a secondary cycle of half the magnitude of the primary cycle:

$$n = \left(\frac{2}{1}\right)^{3.2} = 9.2$$

so it would require 9 secondary cycles to produce the same effect as one primary cycle. If they are occurring at the same frequency, the effective frequency of primary cycling is increased by 1.1 for the purpose of design.

EXAMPLE

A sewer rising main has a pump pressure, including static lift and friction losses, of 600 kPa. When the pump starts up, the pressure rises rapidly to 1.1 MPa before decaying exponentially to the static pumping pressure. On pump shut down, the minimum pressure experienced by the system is 100 kPa. On average, the pump will start up 27 times per day. A minimum life of 100 years is required.

The maximum pressure experienced indicates that a minimum class of PN12.5 will be required. A fatigue analysis is now needed in order to determine suitability or otherwise of PN12.5.

In this system, the pressure range is 1MPa. The pump will start up approximately 1,000,000 times in a 100 year lifetime. However, the exponential cycle pattern means that this should be doubled for design purposes. Therefore, the system should be designed to withstand approximately 2,000,000 cycles in a 100 year lifetime.

Using Figure B.1 to determine the fatigue load factors for polyethylene pipes at 2×10^6 cycles gives the following class selections:

Material	Load factor (Figure B.1)	Equiv. operating pressure (MPa)	Minimum pipe class selection
PE80B or 100	0.8	1.0/0.8 = 1.25	PN12.5
PE80C	0.6	1.0/0.6 = 1.67	N/A

The fatigue analysis thus determines that PN12.5 is suitable for PE80B and PE100, but PE80C may not be used for 100 year life.

EFFECT OF TEMPERATURE

Few data have been established concerning the relationships between fatigue performance and temperature. Joseph (1984) suggests that effects are marginal if any within the range of normal pipeline operating temperatures. It might be expected that higher temperatures could actually inhibit fatigue crack growth by blunting of the crack tip through the reduced yield strength.

Normal temperature de-rating principles should be applied in class selection for maximum operating pressure. Fatigue should then be considered separately and the highest pressure class selected. That is select the highest class arrived via:

A. Static design including temperature derating; or
B. Dynamic design as covered herein.

SAFETY FACTORS

The fatigue design guidelines presented in Table B.1 are based on the lower bound of laboratory test data. They are therefore considered conservative and no additional safety factor need be applied in general. However, where the magnitude or frequency of dynamic stresses cannot be estimated in design with any reasonable degree of accuracy, appropriate caution should obviously be applied. This judgment is in the hands of the designer.

Whilst it is always possible to predict the steady operating conditions with good accuracy, it will occasionally be the case, in complex systems, that it is impossible to predict the extent of surge pressures. In such circumstances, relatively low cost surge mitigation techniques, for example solid state soft-start motor controllers, should be considered. It is of course recommended that actual operating conditions for all systems should be measured, as a matter of routine, when the system is commissioned. Should surge pressure amplitudes in the event exceed expected levels, it is relatively easy matter to retrofit control equipment to ensure that they are kept in check.

DESIGN HINTS

To reduce the effect of dynamic fatigue in an installation, the designer can:

- Limit the number of cycles by:
  - Increasing well capacity for a sewer pumping station;
  - Matching pump performance to tank size to eliminate short demand cycles for an automatic pressure unit; or
  - Using double-acting float valves or limiting starts on the pump by the use of a time clock when filling a reservoir.
- Reduce the dynamic range by:
  - Eliminating excessive water hammer; or
  - Using a larger bore pipe to reduce friction losses.

FITTINGS

With respect to surge and fatigue, fittings performance is dependent upon several factors, and PIPA (2002b) has provided *Part 2 Fusion Fittings for Use with Polyethylene Pressure Pipes Design for Dynamic Stresses* in document POP010B.

APPENDIX C
PROGRAM LISTING FOR PELIFETIME.PY

The Monte Carlo simulation for buried PE pipes has been implemented using the PYTHON 2.3 programming language. PYTHON is freeware, and can be installed from the Internet (www.python.org). For this particular study, PYTHON has allowed complex failure models to be combined with statistical analysis, while maintaining an emphasis on the physical aspects of the problem rather than computing science issues. The full listing for the program PELifetime.py is given below. Text that is proceeded by "#" or "#*" is descriptive, and not part of the PYTHON code.

```
import math
import random
import os
import sys
import string

_root = 'O://PythonPEOutput'  # Comment: Sets directory in which output files are written

#**SECTION 1 - FUNCTION DEFINITIONS FOR FRACTURE MECHANICS MODEL*******

def getSip(p,ro,ri): # Comment: circumferential stress from internal pressure
    val = p*(ro**2+ri**2)/(ro**2-ri**2)
    return val

def getSipRef(p,ri,ro,adef): # Comment: reference stress for internal pressure
    val = -p/(math.log((ri+adef)/ro))
    return val

def getE(E0,t,c): # Comment: Time dependent Young's Modulus in plane strain
    val = E0*(t**(-c))
    return val

def getb(ro,ri): # Comment: Pipe wall thickness
    val = ro-ri
    return val

def getDm(ri,b): # Comment: Pipe mean diameter
    val = 2*(ri+b/2)
    return val

def getYp(adef,b): # Comment: LEFM geometry factor for internal pressure
    val = 1.824-4.669*(adef/b)+(14.959*(adef/b)**2)-(6.341*(adef/b)**3)
    return val

def getKp(Sip,Yp,adef): # Comment: LEFM applied SIF for internal pressure
    val = Sip*Yp*(3.14159*adef)**0.5
    return val

def getSigmaY(sigma0,t,q): # Comment: Time dependent Yield stress
```

```
    val = sigma0*(t**(-q))
    return val

def getSigmaNS(sigmaY,adef,b): # Comment: Net section stress
    val = sigmaY*(1-(adef/b))
    return val

def getSigmaC(Acraze,t,mcraze): # Comment: Craze failure stress
    val = Acraze*(t**(-mcraze))
    return val

def getYd(adef,b): # Comment: LEFM geometry factor for diametrical deflection (From PVC project)
    val = 0.803+0.353*(adef/b)-(1.640*(adef/b)**2)-(1.003*(adef/b)**3)
    return val

def getYr(adef,b): # Comment: LEFM geometry factor for residual stress (from PVC project)
    val = 0.727+2.791*(adef/b)-(2.214*(adef/b)**2)-(2.149*(adef/b)**3)
    return val

def getI(b):  # Comment: Second moment of area for pipe
    val = (b**3)/12
    return val

def getDeflect(E,I,Dm,Esoil,gamma,H, Psurf):  # Comment: Diametrical deflection
    val = 0.0001*Dm*((gamma*H)+Psurf)/((8*E*I/Dm**3)+0.061*Esoil)
    return val

def getSid(E,b,deflect,Dm):  # Comment: Circumferential stress from diametrical deflection
    val = 2*E*b*deflect/(0.1488*(Dm**2))
    return val

def getSidRef(deflect,E,b,adef,Dm):  # Comment: get reference stress for deflection
    val = deflect*E*(b-adef)/(0.584*(Dm**2))
    return val

def getKd(Sid,Yd,adef): # Comment: LEFM applied SIF for diametrical deflection (from PVC project)
    val = Sid*Yd*(3.14159*adef)**0.5
    return val

def getKr(Sir,Yr,adef): # Comment: LEFM applied SIF for residual stress (from PVC project)
    val = Sir*Yr*(3.14159*adef)**0.5
    return val

def interpolateF(a,b,ri): # Comment: Linear interpolation to estimate F from EPRI solution
    F_b_ri005 = (11.947*(a/b)**3) - (4.48*(a/b)**2) + (2.3733*(a/b)) + 0.95 # F function for b/ri = 0.05
    F_b_ri01 = (4.6507*(a/b)**3) + (0.624*(a/b)**2) + (1.1733*(a/b)) + 1.034 # F function for b/ri = 0.1
    F_b_ri02 = (0.3413*(a/b)**3) + (3.328*(a/b)**2) + (0.2347*(a/b)) + 1.108 # F function for b/ri = 0.2
    val = ((F_b_ri005 - F_b_ri02)*((b/ri)-0.2)/(0.05-0.2)) + F_b_ri02
    return val

def interpolateh1(n,a,b,ri):  # Comment: Linear interpolation to estimate h1 from EPRI solution
    h1b_ri005a_b0125 = (0.0058*n**3) - (0.1803*n**2) + (1.693*n) + 3.0148  # h1 function for b/ri = 0.05
        and a/b = 0.125
```

```
    h1b_ri005a_b025 = (0.0153*n**3) - (0.3228*n**2) + (1.9533*n) + 4.0113  # h1 function for b/ri = 0.05
        and a/b = 0.25
    h1b_ri005a_b05 = (-0.0143*n**4) + (0.35*n**3) - (2.8556*n**2) + (7.5168*n) + 5.484  # h1 function
        for b/ri = 0.05 and a/b = 0.5
    h1b_ri005a_b075 = (0.0097*n**4) - (0.3181*n**3) + (3.8232*n**2) - (19.977*n) + 39.69  # h1 function
        for b/ri = 0.05 and a/b = 0.75

    h1b_ri01a_b0125 = (0.0083*n**3) - (0.2149*n**2) + (1.9231*n) + 3.5331  # h1 function for b/ri = 0.1
        and a/b = 0.125
    h1b_ri01a_b025 = (0.0148*n**3) - (0.3189*n**2) + (1.9479*n) + 4.5982  # h1 function for b/ri = 0.1
        and a/b = 0.25
    h1b_ri01a_b05 = (-0.0144*n**4) + (0.3746*n**3) - (3.2917*n**2) + (9.7791*n) + 3.8852  # h1 function
        for b/ri = 0.1 and a/b = 0.5
    h1b_ri01a_b075 = (0.0136*n**4) - (0.3752*n**3) + (3.799*n**2) - (16.956*n) + 29.634  # h1 function
        for b/ri = 0.1 and a/b = 0.75

    h1b_ri02a_b0125 = (0.0047*n**3) - (0.1426*n**2) + (2.0067*n) + 4.4515  # h1 function for b/ri = 0.2
        and a/b = 0.125
    h1b_ri02a_b025 = (0.012*n**3) - (0.2707*n**2) + (1.9114*n) + 5.4038  # h1 function for b/ri = 0.2
        and a/b = 0.25
    h1b_ri02a_b05 = (-0.0116*n**4) + (0.2785*n**3) - (2.1972*n**2) + (5.3021*n) + 6.4387  # h1 function
        for b/ri = 0.2 and a/b = 0.5
    h1b_ri02a_b075 = (0.0108*n**4) - (0.2875*n**3) + (2.7888*n**2) - (11.941*n) + 20.421  # h1 function
        for b/ri = 0.2 and a/b = 0.75

    val1 = ((((a/b) - 0.125)/(0.75 - 0.125)) * (h1b_ri005a_b075 - h1b_ri005a_b0125)) + h1b_ri005a_b0125
        # h1 interpolation for b/ri = 0.05 and actual a/b
    val2 = ((((a/b) - 0.125)/(0.75 - 0.125)) * (h1b_ri01a_b075 - h1b_ri01a_b0125)) + h1b_ri01a_b0125
        # h1 interpolation for b/ri = 0.1 and actual a/b
    val3 = ((((a/b) - 0.125)/(0.75 - 0.125)) * (h1b_ri02a_b075 - h1b_ri02a_b0125)) + h1b_ri02a_b0125
        # h1 interpolation for b/ri = 0.2 and actual a/b

    val4 = ((((b/ri) - 0.05)/(0.2 - 0.05)) * (val3 - val1)) + val1 # h1 interpolation for actual b/ri and actual a/b
    return val4

def getf1(a,ro,ri,F): #Commment:f1 geometry factor for EPFM
    val = (F**2)*12.566*a*(((ro**2)/((ro**2) - (ri**2)))**2)
    return val

def getc(a,ro,ri): #Commment:remaining uncracked ligament
    val = ro-ri-a
    return val

def getp0(c,sigma0,ri,a): #Commment:lower bound limit pressure for perfrect plasticity (EPRI solution)
    val = 1.1547*c*sigma0/(ri+a)
    return val

def getKp(p,ro,ri,a,F): #Commment:Elastic SIF for internal pressure only
    val = 2*p*(ro**2)*((3.14159*a)**0.5)*F/((ro**2) - (ri**2))
    return val

def getKapp(Kp,Kd,Kr): # Comment: Total applied SIF
    val = Kp+Kd+Kr
    return val
```

```
def getPhi(p,po): # Comment:  Crack tip plastic zone multiplier
   val = 1/(1 + ((p/po)**2))
   return val

def getry(n,Kapp,sigma0): # Comment:  Crack tip plastic zone length
   val = 0.0531*((n-1)/(n+1))*((Kapp/sigma0)**2)
   return val

def getaeff(a,phi,ry): # Comment:  Effective crack length
   val = a + (phi*ry)
   return val

def getJp(f1,p,E,alpha,epsilon0,sigma0,c,a,b,h1,p0,n):  # Comment:  Calculate plastic crack driving force per unit
crack width Jp (in MN/m)
   val1 = f1*(p**2)/E
   val2 = alpha*epsilon0*sigma0*c*(a/b)
   val3 = h1*((p/p0)**(n+1))
   val = val1 + (val2*val3)
   return val

def convertJ(J):  # Comment:  Convert units of Jp to N/mm
   val = (J*1e6)/1e3
   return val

def getJe(Kapp,E): # Comment:  Calculate elastoc crack driving force per unit crack width Je (in MN/m)
   val = (Kapp**2)/E
   return val

def getSapp(Sip,Sid,Sir): # Comment: Total applied stress
   val = Sip+Sid+Sir
   return val

#***********************END OF SECTION 1*********************************

#**********************SECTION 2 - INPUT VARIABLES*********************

input = {}
input['Jic'] = 19.0      # Comment:  J0.2 crack initiation toughness for PE material (N/mm)
input['p'] = 0.9     # Comment: Internal Pressure (MPa)
input['ro'] = 177.5E-03        # Comment: Inner radius (Metres)
input['ri'] = 145.2E-03          # Comment: Outer radius (Metres)
input['E0'] = 940.4             # Comment: Short term Young's modulus (MPa) (based on 1 year)
input['m'] = 0.05             # Comment: Viscoelastic parameter for Young's modulus reduction over time (Mod
        in MPa, time in years)
input['adef_shape'] = 1.08   # Comment: Weibull shape parameter for initial defect size (From Barker et al. data
        for 63mm SDR 11 pipes)
input['adef_scale'] = 0.00045   # Comment: Weibull scale parameter for initial defect size (From Barker et al. data
        for 63mm SDR 11 pipes)
input['sigma0'] = 8.7855     # Comment: Short term yield stress (MPa) (based on 1 year)
input['q'] = 0.068           # Comment: Viscoplastic parameter for yield stress reduction over time (stress in MPa,
        time in years)
input['n'] = 1.6      # Comment: Strain hardening exponent
input['alpha'] = 4.7 # Comment: Strain hardening multiplier
```

```
input['epsilon0'] = 0.02 # Comment:  Yield strain
input['Acraze'] = 15.3 # Comment: Craze load bearign parameter
input['mcraze'] = 0.0618 # Comment: Craze damage evolution index
input['Esoil'] = 1.0 # Comment: Soil modulus (MPa)
input['gamma'] = 19           # Comment: Soil unit weight (KN/m^3)
input['H'] = 0.5          # Comment: Soil cover depth (Metres)
input['Psurf'] = 30       # Comment: Surface loading (KPa)
input['Sir'] = 3.3        # Comment: Residual hoop stress at pipe inner surface (MPa)
input['Num'] = 100  # Comment: Number of pipe segments in simulation


#**********************END OF SECTION 2*********************************

#*********SECTION 3 - MONTE CARLO SIMULATION OF PIPE LIFETIMES**********

def getLifetimes(input):
  T = 10000 # Comment: Time period for simulation #
  time = range(1,T)
  adef_shape = input['adef_shape'] # Comment: assigns Weibull shape parameter for defect distribution #
  adef_scale = input['adef_scale'] # Comment: assigns Weibull scale parameter for defect distribution #
  Jic = input['Jic']
  p = input['p']
  ro = input['ro']
  ri = input['ri']
  E0 = input['E0']
  m = input['m']
  Esoil = input['Esoil']
  gamma = input['gamma']
  H = input['H']
  Psurf = input['Psurf']
  Sir = input['Sir']
  Num = input['Num']
  sigma0 = input['sigma0']
  epsilon0 = input['epsilon0']
  Acraze = input['Acraze']
  mcraze = input['mcraze']
  alpha = input['alpha']
  q = input['q']
  n = input['n']
  Sip = getSip(p,ro,ri) # Comment: Calculates internal pressure for input variables #
  b = getb(ro,ri)
  Dm = getDm(ri,b)
  I = getI(b)
  pipes = range(Num) # Comment: Number of pipe segments in simulation #
  Failed = { }         # Comment: Set up dictionary for all pipes #
  Replaced = { }       # Comment: Set up dictionary for replaced pipes #
  NewPipeLog = { }       # Comment: Set up dictionary to record new pipe segments installed over time #

  for id in pipes:
    Failed[id] = [0]  # Comment: Initialise dictionary of failed pipes #
    Replaced[id] = [0] # Comment: Initialise dictionary of replaced pipes #

  for t in time:
      NewPipeLog[t] = []  # Comment: Initialise list of new pipe installation years #
```

```
for i in pipes:
  # Defect size is assigned for 110 SDR 11 pipe from Barker et al. (1983) #
  adef1 = random.weibullvariate(adef_scale,adef_shape) # Comment: assign 1st segment area inherent
    defect sizes to all pipe segments #
  adef2 = random.weibullvariate(adef_scale,adef_shape) # Comment: assign 2nd segment area inherent
    defect sizes to all pipe segments #
  adef3 = random.weibullvariate(adef_scale,adef_shape) # Comment: assign 2nd segment area inherent
    defect sizes to all pipe segments #
  adef = max(adef1,adef2,adef3)
  #adef =  2.2e-3
  for t in time:
    if Replaced[i]:
      adef1 = random.weibullvariate(adef_scale,adef_shape) # Comment: assign 1st segment area inherent
    defect sizes to all pipe segments #
      adef2 = random.weibullvariate(adef_scale,adef_shape) # Comment: assign 2nd segment area inherent
    defect sizes to all pipe segments #
      adef3 = random.weibullvariate(adef_scale,adef_shape) # Comment: assign 2nd segment area inherent
    defect sizes to all pipe segments #
      adef = max(adef1,adef2,adef3)
      #adef = 2.2e-3
    if not Failed[i] or Replaced[i]:
      Yd = getYd(adef,b)
      Yr = getYr(adef,b)
      F = interpolateF(adef,b,ri)    # Calculate LEFM geometry factors based on crack length only
      h1 = interpolateh1(n,adef,b,ri)    # Calculate EPFM geometry factors based on crack length only
      f1 = getf1(adef,ro,ri,F)          # Calculate EPFM geometry factors based on crack length only
      c = getc(adef,ro,ri)              # Calculate remaining ligament length based on crack length only
      sigmaY = getSigmaY(sigma0,t,q)  # Calculate yield stress at time t
      p0 = getp0(c,sigmaY,ri,adef)    # Calculate limit load for plastic collapse based on crack length only
      Kp = getKp(p,ro,ri,adef,F)      # LEFM: Calculate SIF for internal pressure
      E = getE(E0,t,c)                # Calculate Creep modulus at time t
      deflect = getDeflect(E0,I,Dm,Esoil,gamma,H, Psurf)  #  Calculate diametrical deflection
      #print "deflect =", deflect
      Sid = getSid(E0,b,deflect,Dm)   #  Calculate ring bending stress from diametrical deflection
      Kd = getKd(Sid,Yd,adef)    #  LEFM: Calculate SIF from deflection
      Kr = getKr(Sir,Yr,adef)    #  LEFM: Calculate SIF from residual stress
      Kapp = getKapp(Kp,Kd,Kr)   # LEFM: Calculate total applied SIF
      phi = getPhi(p,p0)              # Calculate EPFM plastic zone length factor
      ry = getry(n,Kapp,sigmaY)  # Calculate EPFM plastic zone length
      aeff = getaeff(adef,phi,ry) # Calculate effective crack length including plastic zone length
      F = interpolateF(aeff,b,ri)  # Update LEFM geometry factor based on effective crack length
      f1 = getf1(aeff,ro,ri,F)  # Update EPFM geometry factor based on effective crack length
      h1 = interpolateh1(n,adef,b,ri) # Update EPFM geometry factor based on effective crack length
      Jp = getJp(f1,p,E0,alpha,epsilon0,sigmaY,c,aeff,b,h1,p0,n) # Calculate plastic component of crack
    driving force per unit crack width
      Jp = convertJ(Jp) # Convert Jp to (N/mm)
      Je = getJe(Kapp,E0) # Calculate elastic component of crack driving force per unit crack width
      Je = convertJ(Je) # Convert Je to (N/mm)
      Japp = Jp + Je # Calculate total crack driving force
      #print "deflect =", deflect
      #print "p =", p
      #print "Japp =", Japp
      sigmaNS = getSigmaNS(sigmaY,aeff,b)
```

```
            #print "SigmaNS =", sigmaNS
            #Sapp = getSapp(Sip,Sid,Sir)
            Sapp = Sip + Sid + Sir
            #print "Sapp =", Sapp
            SipRef = getSipRef(p,ri,ro,adef) #  Calculate reference stress for internal pressure
            #print "SipRef =", SipRef
            SidRef = getSidRef(deflect,E0,b,adef,Dm) # Calculate reference stress for deflection
            #print "SidRef =", SidRef
            SrefTot = SipRef + SidRef + Sir
            #print "SRefTot =", SrefTot
            sigmaC = getSigmaC(Acraze,t,mcraze) # Calculate craze failure stress
            #print "sigmaC =", sigmaC
            if Jic <= Japp:  # Comment: Check for elastic plastic crack initiation
                Failed[i] += [t]  # Comment: Add failure year to list for pipe "i" #
                Replaced[i] += [t] # Comment: Add installation year to list for  pipe "i" #
                NewPipeLog[t] += [1] # Comment: Add "1" into new pipe list at time "t" #
            else:
              if aeff >= b: # Comment: Check for effective crack sizes greater than wall thickness  #
                    Failed[i] += [t]
                    Replaced[i] += [t]
                    NewPipeLog[t] += [1]
              else:
                if sigmaNS <= Sapp: # Comment: Check for ductile fracture failure  #
                      Failed[i] += [t]
                      Replaced[i] += [t]
                      NewPipeLog[t] += [1]
                else:
                if sigmaC <= SrefTot: # Comment: Check for craze fialure and crack initiation
                      Failed[i] += [t]
                      Replaced[i] += [t]
                      NewPipeLog[t] += [1]

#***********************END OF SECTION 3 ***********************************

#***********************SECTION 4 - WRITE OUTPUT TO TEXT FILES************

  wrtDict(_root+'//FailHistory.txt','Segment id,[Failure years]',Replaced)
  wrtDict2(_root+'//NewSegment.txt','Year, Number of new segments installed',NewPipeLog)

  r_dict = { }
  r_dict['replaced'] = Replaced
  r_dict['failed'] = Failed
  r_dict['newPipeLog'] = NewPipeLog

  return r_dict

def wrtDict(fname,title,dict): # Comment: Procedure for writing failure years to each pipe segment in the pipe
      dictionary #
  data  = file(fname,'w')
  data.write(title)
  for id in dict.keys():
    line = "%s," %id
    line += "%s" %dict[id]
    data.write(line)
```

```
    data.close()

def wrtDict2(fname,title,dict): # Comment: Procedure for writing number of new installations to the pipe dictionary #
    data  = file(fname,'w')
    data.write(title)
    for id in dict.keys():
        line = "%s," %id
        line += "%s" %len(dict[id])
        data.write(line)
    data.close()

#************************END OF SECTION 4************************************

#************************END OF PROGRAM************************************
```

APPENDIX D
EXAMPLE CALCULATION OF EXPECTED FAILURE RATE IN 100MM (DN 4 INCH) SDR 11 PE PIPES

In chapter 9, an equation for the calculation of estimated values of expected failure rates in 100mm (or DN 4 inch) SDR 11 PE pipes was presented:

$$E(N) = nh(t) \tag{D.1}$$

where n is the number of 1-metre long segments in the pipeline, and $h(t)$ is the hazard/failure rate function (Crowder et al. 1991). In practice, $h(t)$ is a measure of the probability of imminent failure at time t, given that failure has not yet occurred (Crowder et al. 1991). In turn, $h(t)$ is written in terms of a scale parameter α, and the shape parameter η as:

$$h(t) = \eta\alpha^{-\eta}t^{\eta - 1} \tag{D.2}$$

α and η values for different PE materials and operating pressures and are shown in the following table.

Material	Operating pressure in MPa and psi	Weibull scale (α)	Weibull shape (η)
PE80C (AUS)	0.5 (72.5)	1,021.8	3.9
	0.6 (87.0)	779.2	3.9
	0.7 (101.5)	563.5	3.7
	0.8 (116.0)	248.6	3.2
	0.9 (130.5)	110.8	3.2
PE80B (AUS)	0.5 (72.5)	2,594.8	3.9
	0.6 (87.0)	2,016.7	4.0
	0.7 (101.5)	1,455.7	3.7
	0.8 (116.0)	814.7	2.9
	0.9 (130.5)	279.6	3.2
PE100 (AUS)	0.5 (72.5)	8,423.0	3.9
	0.6 (87.0)	6,597.6	3.8
	0.7 (101.5)	5,027.1	2.7
	0.8 (116.0)	2,319.8	3.1
	0.9 (130.5)	1,076.3	2.7
PE3408 (US)	0.5 (72.5)	8,787.6	4.1
	0.6 (87.0)	6,765.1	4.0
	0.7 (101.5)	5,178.9	3.2
	0.8 (116.0)	2,243.1	3.3
	0.9 (130.5)	1,034.0	2.9

To demonstrate how the expected failure rate can be calculated in a PE pipe, consider the hypothetical example of a 110mm SDR 11 PE pipeline, manufactured from PE80B (AUS) material and operating at an internal pressure of 0.8 MPa (116 psi). Let us assume that the expected failure rates (per pipe length/per year) at 50 and 100 years of age are required.

Step 1—*Extract scale and shape parameters from table:*
From the table above, values of $\alpha = 814.7$ and $\eta = 2.9$ are adopted.

Step 2—*Calculate the hazard function from Equation D.2*
From Equation D.2:

At 50 years of age, h(t) = 2.9 * (814.7^–2.9) * (50^(2.9–1)) = 1.77×10^{-5}
At 100 years of age, h(t) = 2.9 * (814.7^–2.9) * (100^(2.9–1)) = 6.61×10^{-5}

Step 3—*Calculate expected failure rate from Equation D.1*
In a 100km long length of pipe, there are 100,000 1-metre long segments. This means that $n = 100{,}000$. Therefore, from Equation D.1:

At 50 years of age, E(N) = 100,000 * 1.77×10^{-5} = 1.77 per 100km/per year.
At 100 years of age E(N) = 100,000 * 6.61×10^{-5} = 6.61 per 100km/per year.

Step 4—*Convert failure rates to events per mile/per year if required*

At 50 years of age, E(N) = 1.77 / 0.6 = 2.83 per 100 mile/per year.
At 100 years of age, E(N) = 6.61 / 0.6 = 10.58 per 100 mile/per year.

REFERENCES

Ahammed, M., and R.E. Melchers. 1997. Probabilistic Analysis of Underground Pipelines Subject to Combined Stress and Corrosion. *Engng. Struct.*, 19(12):988–994.

Allcard, C., and S. Beech. 2001. In *Proc. of the Eleventh International Conference on Plastics Pipes, Munich, Germany.* Plastics and Rubber Institute.

Andersson, U. 2001. In *Proc. of the Eleventh International Conference on Plastics Pipes, Munich, Germany.* Plastics and Rubber Institute.

ANSI/NSF (American National Standards Institute and NSF International). 2001. *Drinking Water System Components,* ANSI/NSF 14. Ann Harbor, Mich.: ANSI.

ANSI/NSF (American National Standards Institute and NSF International). 2002. *Drinking Water System Components—Health Effects,* ANSI/NSF 61. Ann Harbor, Mich.: ANSI.

ASTM International. 2001a. *Standard Test Method for Environmental Stress Cracking of Ethylene Plastics,* ASTM F1693-01. Philadelphia: ASTM International.

ASTM International. 2001b. *Standard Test Method for Notch Tensile Test to Measure the Resistance to Slow Crack Growth of Polyethylene Pipes and Resins,* ASTM F1473-01. Philadelphia: ASTM International.

ASTM International. 2002. *Standard Specification for Polyethylene Plastics Pipe and Fittings Materials,* ASTM D3350-02a. Philadelphia: ASTM International.

ASTM International. 2003. *Standard Test Method for Obtaining Hydrostatic Design Basis for Thermoplastic Pipe Materials or Pressure Design Basis for Thermoplastic Pipe Products,* ASTM D2837-02a. Philadelphia: ASTM International.

Atkins, A.G., and Y.W. Mai. 1988. *Elastic and Plastic Fracture.* Chichester, UK: Ellis Horwood.

AWWA (American Water Works Association). 1999. *Polyethylene (PE) Pressure Pipe and Fittings, 4 in. (100mm) Through 63 in. (1,575mm) for Water Distribution and Transmission.* AWWA C906. Denver, Colo.: AWWA.

Barker, M.B., J. Bowman, and M. Bevis. 1983. The Performance and Causes of Failure of Polyethylene Pipes Subjected to Constant and Fluctuating Internal Pressures. *Jour. Mater. Sci.,* 18:1095–1118.

Beech, S.H. 2006. Private communication.

Beech, S.H., C.R. Ferguson, and E.Q. Clutton. 2001. In *Proc. of the Eleventh International Conference on Plastics Pipes, Munich, Germany.* Plastics and Rubber Institute.

Beech, S.H., A. Headford, S. Hunt, and G. Sandilands. 1995. The Resistance of Polyethylene Water Pipeline Systems to Surge Pressures. In *Proc. of the Plastics Pipes IX Conference, Edinburgh, Scotland.*

Bernal, C., H. Lopez Montenegro, and P. Frontini. 2003. Failure Prediction Analysis for Polyethylene Flawed Pipes. *Engng. Frac. Mech.,* 70:2149.

Berthold, L., L. Bohm, H.F. Enderle, V. Lackner, D. Lilge, and U. Schulte. 2001. In *Proc. of the Eleventh International Conference on Plastics Pipes, Munich, Germany.* Plastics and Rubber Institute.

Boros, S.J. 1995. ASTM vs. ISO Methodology for Pressure Design of Polyethylene Piping Materials. In *Proc. of the Fourteenth International Plastic Fuel Gas Symposium, San Diego, Calif.* American Gas Association.

Bowman, J. 1995. In *Proc. of the Ninth International Conference on Plastics Pipes, Edinburgh, Scotland.* Plastics and Rubber Institute.

Bragaw, C.G. 1978. Crack Stability Under Load and the Bending Resistance of MDPE Piping Systems. In *Proc. of the Sixth Plastic Fuel Gas Pipe Symposium*. American Gas Association.

Bragaw, C.G. 1983. Service Rating of Polyethylene Piping Systems by the Rate Process Method. In *Proc. of the Eighth Plastic Fuel Gas Pipe Symposium, Columbus, Ohio*. American Gas Association.

Broberg, K.B. 1999. Influence of T-Stress, Cohesive Strength and Yield Strength on the Competition Between Decohesion and Plastic Flow in a Crack Edge Vicinity. *Int. Jour. Fract.*, 4:11.

Brown, N., and X. Lu. 1993. Controlling the Quality of PE Gas Piping Systems by Controlling the Quality of the Resin. In *Proc. of the Thirteenth International Plastic Fuel Gas Pipe Symposium, San Antonio, Texas*. American Gas Association.

Brown, N., and X. Lu. 2001. The Dependence of Rapid Crack Propagation in Polyethylene Pipes on the Plane Stress Fracture Energy of the Resin. *Polym. Eng. Sci.*, 41:1140.

BSI (Bristish Standards Institute). 2000. Suitability of Non-metallic Products for Use in Contact with Water Intended for Human Consumption with Regard to Their Effect on the Quality of the Water—Specification, BS 6920-1. London: BSI.

Bucknall, C.B., and P. Dumpleton. 1985a. Factors Affecting Fatigue Crack Growth in HDPE. *Plast., Rubb. Proc. App.*, 5:343–347.

Bucknall, C.B., and P. Dumpleton. 1985b. Fatigue Crack Growth in Polyethylene, *Polym. Eng. Sci.*, 25:343–347.

Burn, L.S., and B.L. Schafer. 1997. The Environmental Impact of Lead Leaching from uPVC Sewerage Waste and Vent Pipes. *Technical Report TR97/1*. Highett, Australia: CSIRO Building, Construction and Engineering.

Burn, S., P. Davis, T. Schiller, B. Tiganis, G. Tjandraatmadja, M. Cardy, S. Gould, and P. Sadler. 2005. *Long-Term Performance Prediction of PVC Pipes* AwwaRF Project No. 2879. Denver, Colo.: AwwaRF.

Chan, M.K.V., and J.G. Williams. 1981. Plane Strain Fracture Toughness Testing of High Density Polyethylene. *Polym. Eng. Sci.*, 21:15:1019–1026.

Chudnovsky, A., D. Baron, and Y. Schulkin. 1995. Lifetime Prediction and Reliability Evaluation for Gas Piping Grade Polyethylene, In *Proc. of the Fourteenth International Plastic Fuel Gas Pipe Symposium, San Diego, Calif.* American Gas Association.

Chudnovsky, A., K. Sehanobish, and S. Wu. 2000. Methodology for Durability Analysis of HDPE Pipe. *Transactions ASME*, 122:152–155.

Coleman, B.D. 1956. *Jour. Polym. Sci.*, XX:447.

Crowder, M.J., A.C. Kimber, R.L. Smith, and T.J. Sweeting. 1991. *Statistical Analysis of Reliability Data*. London: Chapman and Hall.

Davis, P., C.J. Greenshields, V. Dabas, and P.S. Leevers. 1998. In *Proc. of the Tenth International Conference, "Plastics Pipes," Goteborg, Sweden.*

Duan, D., and J.G. Williams. 1998. Craze Testing for Tough Polyethylene. *Jour. Mat. Sci.*, 3:635.

Gaunt, J.A., and S.K. Ong. 2005. Water Utility Experiences with Plastic Pipes and Gaskets in Hydrocarbon Contaminated Soils. Denver, Colo.: AwwaRF.

Greenshields, C.J. 1997. *Plast. Rubber Process. Applic.*, 26:387.

Greenshields, C.J., P. Davis, and P.S. Leevers. 1997. *Plast. Rubber Process. Applic.*, 26:404.

Greenshields, C.J., and P.S. Leevers. 1999. Correlation Between Full Scale and S4 Tests for Rapid Crack Propagation in Plastic Gas Pipe. *Plast. Rubber Compos.*, 28:20.

Greig, J.M. 1988. In *Proc.of the Eighth International Conference, "Plastics Pipes," Bath, England.*

Greig, J.M. 1996. *Plast. Rubber Process. Applic.*, 25:163.

Hale, G.E., and F. Ramsteiner. 2000. J-Fracture Toughness of Polymers at Slow Speed. In *ESIS-TC4 Polymers and Composites*. Cambridge, UK.

HBN (Healthy Building Network) 2005 [Online]. Available: http://www.healthybuilding.net/pvc/seattlepipes.html [cited Feb 13, 2006]

Hill, D.J., K.A. Wilson and A.L. Headford. 2001. In *Proc. of the Eleventh International Conference on Plastics Pipes, Munich, Germany.* Plastics and Rubber Institute

ISO (International Organization for Standardization). 1995. *Thermoplastics Materials for Pipes and Fittings for Pressure Applications Classification and Designation—Overall Service (Design) Coefficient.* ISO 12162. Geneva: ISO.

ISO (International Organization for Standardization). 1997. *Polyolefin Pipes for the Conveyance of Fluids—Determination of Resistance to Crack Propagation. Test Method for Slow Crack Growth on Notched Pipes (Notch Test).* ISO 13479. Geneva: ISO.

ISO (International Organization for Standardization). 2001. *Plastic Piping Systems for the Transport of Water Intended for Human Consumption: Migration Assessment—Determination of Migration Values for Plastic Pipes and Fittings and Their Joints.* ISO 8795. Geneva: ISO.

ISO (International Organization for Standardization). 2003. *Plastics Pipes and Ducting Systems—Determination of the Long-Term Hydrostatic Strength of Thermoplastics Materials in Pipe Form by Extrapolation.* ISO 9080. Geneva: ISO.

ISO (International Organization for Standardization). 2004. *Plastics—Determination Environmental Stress Cracking of Polyethylene—Full Notch Creep Test.* ISO 16770. Geneva: ISO.

ISO (International Organization for Standardization). 2005. *Plastics Piping Systems—Polyethylene (PE) Pipes and Fittings for Water Supply—Part 1: General.* ISO 4427-1. Geneva: ISO.

Janson, L.E. 1995. *Plastic Pipes for Water Supply and Sewage Disposal.* Borealis, Stockholm.

Jarrett, R., O. Hussain, A. Veevers, and J. Van der Touw. 2001. In *Proc. of the International Conference of Maintenance Societies, Melbourne, Australia.*

Joseph, S.H. 1979. The Pressure Fatigue Testing of Plastic Pipes. In *Plastics Pipes 4, PRI, London*, Paper 28.

Joseph, S.H. 1984. Fatigue Failure and Service Lifetime in uPVC Pressure Pipes. *Plast., Rubber Process. App.*, 4(4):325–330.

Kanninen, M.F., C.J. Kuhlman, and M.M. Mahmoun. 1993. Procedures for Forecasting Pipe Life in Utility Service Installations. In *Proc. of the Thirteenth International Plastic Fuel Gas Pipe Symposium, San Antonio, Texas.* American Gas Association.

Kanninen, M.F., P.E. O'Donoghue, C.H. Popelar, and V.H. Kenner. 1990. A Visco-Elastic Fracture Mechanics Assessment of Slow Crack Growth in Polyethylene Gas Distribution Pipe Materials. *Eng. Fract. Mech.*, 36(6):903–918.

Kuhlman, C.J., G.G. Chell, C.H. Popelar, R.J. Lewis, T.S. Grant, M.M. Mamoun, D.G. Butts, D.D. Skerhut, and S.E. Stewart. 1997. PE Lifespan Forecasting: A Computer Program for Evaluating the Design of Directional Drilling Profiles and Polyethylene Pipelines for Long-Term Service. In *Proc. of the International Plastic Pipe Symposium, Orlando, Fla.*

Kumar, V., M.D. German, and C.F. Shih. 1981. *An Engineering Approach for Elastic Plastic Fracture Analysis*. EPRI Report NP-1931. Palo Alto, Calif.: EPRI.

Kuraoka, S., B. Rajani, and C. Zhan. 1996. Pipe-Soil Interaction Analysis of Field Tests of Buried PVC Pipe. *Jour. Infrastruct. Sys.*, 2(4):162–170.

Leevers, P.S. 1995. *Int. Jour. Fract.*, 73:109.

Leevers, P.S., L. Moreno, and A. Paizis. 2001. In *Proc. of the Eleventh International Conference on Plastics Pipes, Munich, Germany*. Plastics and Rubber Institute.

Lin, L., and A.S. Argon. 1994. Review: Structure and Plastic Deformation in Polyethylene. *Jour. Mater. Sci.*, 29:294–323.

Lu, J.P., P. Davis, and L.S. Burn. 2004. Lifetime Predictions for ABS Pipes Subjected to Combined Pressure and Deflection Loading. *Polym. Eng. Sci.*, 43:444.

MacKellar, S., and D. Pearson. 2003. Nationally Agreed Failure Data and Analysis Methodology for Water Mains Volume 1: Overview and Findings. Report Ref. No. 03/RG/05/7, UK Water Industry Research. London: UKWIR.

Marshall, G.P. 1982. Design for Toughness in Polymers: 1—Fracture Mechanics. *Plast., Rubber Compos. Process. Appl.*, 2(2):169–182.

Marshall, G.P., and S. Brogden. 1997. Evaluation of Long Term Performance—The Behaviour of Buried Pipes. Report Ref. No. 97/WM/02/5, UK Water Industry Research. London: UKWIR.

Marshall, G.P., S. Brogden, and M.A. Shepherd. 1998. Evaluation of the Surge and Fatigue Resistance of PVC and PE Pipeline Materials for Use on the U.K. Water Industry. In *Proc. of the Plastics Pipes X Conference, Gothenburg, Sweden.*

Marshall, G.P., D. Pearson, and S. MacKellar. 1996. Specification of Performance for Modified Poly(Vinyl Chloride) and Polyethylene Pipes. *Plast., Rubber Compos. Process. Appl.*, 25(6):276–286.

Melchers, R.E. 1999. *Structural Reliability Analysis and Prediction*. 2nd ed. Brisbane, Australia: John Wiley and Sons.

Mruk, S. 2006. Private communication.

Nagy, K., E. Epacher, P. Staniek, and B. Pukánszky. 2003. Hydrolytic Stability of Phenolic Antioxidants and Its Effect on Their Performance in High-Density Polyethylene. *Pol. Deg. Stab.*, 82:211–219.

Palermo, E.F., and I.K. DeBlieu. 1985. Rate Process Concepts Applied to Hydrostatically Rating Polyethylene Pipe. In *Proc. of the Ninth Plastic Fuel Gas Pipe Symposium, New Orleans, La.*

Palermo, G. 2004. Correlating Aldyl "A" and Century PE Pipe Rate Process Method Projections with Actual Field Performance. In *Proc. of the Twelfth International Conference, "Plastics Pipes," Milan, Italy.*

Pandya, K.C., and J.G. Williams. 2000. Measurement of Cohesive Zone Parameters in Tough Polyethylene. *Polym. Eng. Sci.*, 40:1765.

Parmar, R., and J. Bowman. 1989. *Polym. Engng. Sci.*, 29:19.

Parsons, M., E.V. Stepanov, A. Hiltner, and E. Baer. 2000. Correlation of Fatigue and Creep Slow Crack Growth in a Medium Density Polyethylene Pipe Material. *Jour. Mat. Sci.*, 35:2674.

PIPA (Plastics Industry Pipe Association of Australia, Ltd.). 2002a. *Polyethylene Pressure Pipes Design for Dynamic Stresses, Issue 5.0. Ref POP010A*. Sydney, Australia: PIPA.

PIPA (Plastics Industry Pipe Association of Australia, Ltd.). 2002b. *Fusion Fittings for use with Polyethylene Pressure Pipes Design for Dynamic Stresses, Issue 5.0. Ref POP010B*. Sydney, Australia: PIPA.

Plummer, C.J.G., A. Goldberg, and A. Ghanem. 2001. Micromechanisms of Slow Crack Growth in Polyethylene Under Constant Tensile Loading. *Polymer,* 42:9551.

Popelar, C.F., C.H. Popelar, and V.H. Kenner. 1990. Viscoelastic Material Characterization and Modeling for Polyethylene. *Poly. Eng. Sci.,* 30:577–586.

Popelar, C.H. 1993. A Comparison of the Rate Process Method and the Bidirectional Shifting Method. In *Proc. of the Thirteenth International Plastic Fuel Gas Pipe Symposium, San Antonio, Texas*. American Gas Association.

PPI (Plastics Pipe Institute). 2007. *Handbook of PE Pipe* [Online]. Available: <http://plasticpipe.org/publications/pe_handbook.html>. [cited October 31, 2006]

Rajani, B., S. McDonald, and G. Felio. 1995. Water Mains Break Data on Different Pipe Materials for 1992 and 1993. Ottawa, Ontario: National Research Council of Canada.

Rajani, B., and C. Zhan. 1996. On the Estimation of Frost Loads. *Can. Geotech. Jour.*, 33:629–641.

Ritchie, S.J.K., P. Davis, and P.S. Leevers. 1998. Brittle Tough Transition of Rapid Crack Propagation in Polyethyelene. *Polymer.*, 39:6657–6663.

Rooke D.P., and Cartwright, D.J. 1976. *Compendium of Stress Intensity Factors*. London: HMSO.

Rubeiz, C.G. 2004 Case Studies on the Use of HDPE Pipe for Municipal and Industrial Projects in N. America. In *Proc. of the Twelfth International Conference, "Plastics Pipes," Milan, Italy.*

Sandilands, G.J., and J. Bowman. 1986. An Examination of the Role of Flaw Size and Material Toughness in the Brittle Fracture of Polyethylene Pipes. *Jour. Mater. Sci.,* 21:2881–2888.

Sandstrum, S.D. 2004. High Density Polyethylene (HDPE) water distribution systems. In *Proc. of the Twelfth International Conference, "Plastics Pipes," Milan, Italy.*

Schulte, C.A. 1960. In *ASTM STP60 (American Society for Testing and Materials)*. Philadelphia.

Sehanobish, K., A. Moet, A. Chudnovsky, and P.P. Petro. 1985. Fractographic Analysis of Field Failure in Polyethylene Pipe. *Jour. Mater. Sci. Lett.,* 4:890–894.

Shah, A., E.V. Stepanov, G. Capaccio, A. Hiltner, and E. Baer. 1998. Stepwise Fatigue Crack Propagation in Polyethylene Resins of Different Molecular Structure. *Jour. Polym. Sci., Part B: Polym. Phys.,* 36:2355–2369.

Stahmer, M. 2006. Private communication.

Standards Australia. 1988. *Polyethylene Pipe Extrusion Compounds*, AS 1463. Sydney, Australia: Standards Australia.

Standards Australia. 2001. *Determination of Tensile Properties of Plastics Materials—Test Conditions for Moulding and Extrusion Plastics*, AS 1145.2. Sydney, Australia: Standards Australia.

Standards Australia/Standards New Zealand. 1998. *Buried Flexible Pipelines Part 1: Structural Design,* AS/NZS 2566.1. Sydney, Australia: Standards Australia.

Standards Australia/Standards New Zealand. 2000. *Products for Use in Contact with Drinking Water*, AS/NZS 4020. Sydney, Australia: Standards Australia.

Standards Australia/Standards New Zealand. 2003a. *Polyethylene (PE) Compounds for Pressure Pipes and Fittings,* AS/NZS 4131. Sydney, Australia: Standards Australia.

Standards Australia/Standards New Zealand. 2003b. *Polyethylene (PE) Pipes for Pressure Pipes and Fittings,* AS/NZS 4130. Sydney, Australia: Standards Australia.

Strebel, J.J., and A. Moet. 1991. Accelerated Fatigue Fracture Mechanism of Medium Density Polyethylene Pipe Material. *Jour. Mat. Sci.*, 26:5671–5680.

Thoft-Christensen, P., and M.J. Baker. 1982. *Structural Reliability Theory and Its Applications.* Berlin: Springer-Verlag.

van Driel, M. 1997. A New Methodology for Quality Assessment and Lifetime Prediction of Gas Distribution Mains. In *Proc. of the International Plastic Pipe Symposium, Orlando, Fla.*

Wang, X. 2002. Elastic T-Stress for Cracks in Test Specimens Subjected to Non-Uniform Stress Distributions. *Engng. Fra.c Mech.*, 69:1339.

Wang, X., and R. Bell. 2004. Elastic T-Stress Solutions for Semi-Elliptical Surface Cracks in Finite Thickness Plates Subject to Non-Uniform Stress Distributions. *Engng. Frac. Mech.*, 71:1477.

WHO (World Health Organization). 1993. *Guidelines for Drinking-Water Quality*: *Vol. 1—Recommendations.* 2nd ed. [Online]. Available: <http://www.who.int/ water_sanitation_health/GDWQ/Chemicals/leadsum.htm>. [cited October 20, 2006]

WHO (World Health Organization). 1996. *Guidelines for Drinking-Water Quality: Vol. 2—Health Criteria and Other Supporting Information* [Online]. Available: <http://www.who.int/water_sanitation_health/GDWQ/Summary_Tables/Tab2d.htm>. [cited November 15, 2006]

Williams, J.G. 1978. Applications of Linear Fracture Mechanics. *Adv. Polym. Sci.*, 27:67–120.

Williams, J.G. 2000. Introduction to Elastic Plastic Fracture Mechanics. In *ESIS-TC4 Polymers and Composites.* Cambridge, UK.

Yayla, P., and P.S. Leevers. 1992. Rapid Crack Propagation in Pressurised Plastic Pipe—II. Critical Pressures for Polyethylene Pipe. *Engng. Frac. Mech.*, 42:675.

Zhou, Y., and N. Brown. 1993. Evaluating the Fatigue Resistance of Notched Specimens of Polyethylene. *Polym. Eng. Sci.*, 33:1421.

ABBREVIATIONS

a	crack size
AS	Australian Standards
ASTM	American Society for Testing and Materials or ASTM International
AWWA	American Water Works Association
AwwaRF	Awwa Research Foundation
b	wall thickness (mm)
BSI	British Standards Institute
°C	degrees Celsius
CDNT	Circumferentially Deep Notched Tensile
CEN	European Commission Norm
COD	crack opening displacement
CSIRO	Commonwealth Scientific and Industrial Research Organisation
DENT	double edge notch tensile
DSC	differential scanning calorimetry
δ^*	critical crack opening displacement
Δ	delta or change in
E	Young's modulus
EDX	energy dispersive X-ray spectroscopy
EF	electrofusion
EN	European Standard
EPA	Environmental Protection Agency
EPFM	elastic plastic fracture mechanics
ESIS	European Structural Integrity Society
EWF	essential work of fracture
F	safety factor
FCNT	full notch creep test
FEA	finite element analysis
FOSM	first-order second-moment
HALS	hindered amine light stabilizers
HDB	hydrostatic design basis
HDPE	high density polyethylene
HDS	hydrostatic design stress
ISO	International Organization for Standardization
J	crack resistance
K_I	applied stress intensity factor (SIF)
K_{IC}	fracture toughness
K_{ID}	applied SIF from diametrical deflection

K_{IP}	applied SIF from internal pressure
K_{IR}	applied SIF from residual stress
K_{ITOT}	total applied SIF
LCL	lower confidence limit
LDIW	low ductile inner wall
LDPE	low density polyethylene
LEFM	linear elastic fracture mechanics
LIFESPAN	Life Integrity Forecasting and Enhanced Service Performance ANalysis
LTHS	long-term hydrostatic design stress
MAL	maximum allowable level
MCL	maximum contaminant level
MDPE	medium density polyethylene
MPa	megapascals
MRS	minimum required strength
NPT	notched pipe test
NSF	National Sanitation Foundation
OIT	oxygen induction time
PE	polyethylene
PENT	Pennsylvania notch test
PPI	Plastics Pipe Institute
RCP	rapid crack propagation
RPM	rate process method
SCB	short chain branch
SCG	slow crack growth
SEM	scanning electron microscope/microscopy
SENB	single edge notch bend
SIF	stress intensity factor
σ	remotely applied stress
σ_Y	yield stress
3PB	three point bend
t_i	incubation time for crack initiation
UK	United Kingdom
UKWIR	UK Water Industry Research
USEPA	US Environmental Protection Agency
WHO	World Health Organization
WRc	UK Water Research Centre
WSAA	Water Services Association of Australia
Y	fracture mechanics geometry factor

IWA Publishing's authorised EU representative for General Product Safety Regulations is Diane D'Arras, 15 rue Duret, 75116 Paris, France, e-mail: safety@iwap.co.uk.

Printed and bound by CPI Group (UK) Ltd, Croydon, CR0 4YY
17/09/2026
02218793-0001